The Lost Dutchman Mine of Jacob Waltz Part II

The Holmes Manuscript

by
George "Brownie" Holmes

Compiled and Annotated by
T.E. Glover, PhD

First Edition

The Holmes Manuscript

Part II: The Lost Dutchman Mine of Jacob Waltz

Compiled and Annotated by TE Glover, PhD

Published by: Cowboy Miner Productions

Cowboy-Miner Productions,
PO Box 9674
Phoenix, Arizona, 85068

First Printing

Printed and bound in the United States

Cover design by ATG Productions, Phoenix, Arizona
Photographs: Front cover insert: Brownie Holmes holding the skull of Adolph Ruth photographed by Ed Newcomer (Nyles Latham Collection). Rear cover insert: Richard Holmes courtesy of Holmes family; Gideon Roberts courtesy of the Roberts family.

Photographs not otherwise credited were taken by the author.

Publishers Cataloging-in-Publication Data:

Glover, Thomas E.

The Homes Manuscript, Part II The Lost Dutchman Mine of Jacob Waltz

1st ed.

176 p. : ill : maps

Includes index

1. Lost Dutchman Mine — Treasure-trove — Superstition Mountains — Arizona. 2. George "Brownie" Holmes — Biography — Holmes family; 3. Superstition Mountains (Ariz.) — History. 4. Superstition Mountains — Description — Travel. 5. Treasure-trove — Arizona.

979.1 G 99-068881

0-9662091-8-4 (softcover) : $17.95

10 9 8 7 6 5 4 3 2 1

Dedicated to
Geroge "Brownie" Holmes
and
Billy Harwood

Warning and Disclaimer

Hiking, backpacking, horseback riding, climbing and all outdoor activities are inherently dangerous. Off trail hiking and looking for the Lost Dutchman Mine present additional risk. The information provided in this book is intended to entertain and inform. The final judgement and decision to pursue any outdoor activity is always your responsibility as the user of this book.

It is your responsibility to obtain the skills, abilities and knowledge necessary to pursue any search for the mine, or engage in horseback riding, hiking, climbing and other outdoor activities. You must be physically fit, and know about survival, flash floods, topography, cactus, snakes, scorpions and other wildlife before attempting active searching. Terrain often changes due to flash floods, erosion and other natural or man made events. Although I have made considerable effort to test the accuracy of the information in this book, errors and omissions most certainly occur. You must decide whether or not to search a particular area, whether or not the weather and ground conditions are safe, the manner of the search and whether or not to initiate or continue a search. You must decide whether you possess the skills, abilities and fitness required to pursue a particular trip.

The authors, publishers and those associated with this publication, directly or indirectly, assume no responsibility for any accident, injury, damage or loss that may result from participating in any activities described, referred to or inferred in this book. This is a guide to the legend and history as best as I have been able to understand it — in searching you are on your own, and your decisions are totally your responsibilty.

Table of Contents

Foreword

This account of The Lost Dutchman Mine and Jacob Waltz has traditionally been called *The Holmes Manuscript.* It is believed to be the only surviving pioneer account of any of the families originally involved with Waltz and his mine. Dick Holmes and Gideon Roberts were the only two persons with Jacob Waltz when Waltz died. It was to Holmes and Roberts that Waltz made his deathbed confession, telling the story of the Lost Dutchman Mine, and it was to them that Waltz revealed the directions to the mine. Over 40 years after Dick Holmes' death, a manuscript reportedly written by Dick Holmes' son, George "Brownie" Holmes, appeared somewhat mysteriously at the Arizona State Library. It arrived without pedigree, as one of three manuscripts deposited by an Arizona newspaperman, Charles F. Higham, who had a bit of a reputation for tall tales. Was *The Holmes Manuscript* genuine or someone's idea of a hoax? After all, fakes in the lore and history of the Lost Dutchman Mine are not unknown. In my researches, I had encountered many such fakes, ranging from simple maps to a handwritten "original" manuscript of Sims Ely's entire book *The Lost Dutchman Mine.*

At first inspection, the matter appeared a simple open and shut case: *The Manuscript* was a fake. Why? Because the man who had deposited it at the Arizona State Library, Charles F. Higham, denounced it and the man who reportedly wrote it, George "Brownie" Holmes, disowned it. Higham's remarks were contained in a letter from Higham denouncing *The Manuscript* as worthless at best and possibly a fake. This letter accompanied *The Manuscript* when Higham donated it to the Arizona State Library.

The clincher, however, appeared to be Brownie Holmes' own remarks about *The Manuscript.* Swanson and Kollenborn reported:

> At the Arizona Archives [sic. State Library] there is a manuscript, allegedly written by "Brownie" Holmes, which was placed there by Charles F. Higham. When "Brownie" Holmes examined the manuscript, he stated that he had never seen it before. (*Superstition Mountain, A Ride Through Time*, 1981, Ch. 14, p. 121)

Investigating the lore and history of the Lost Dutchman Mine, I originally discounted *The Manuscript* as one of the many fanciful accounts of the story. But, as I delved deeper into the history of the Lost Dutchman Mine, a strange thing happened. While other versions of the story seemed to collapse when subjected to scrutiny, *The Holmes Manuscript* slowly gained credibility. The tale told in *The Manuscript* was a much simpler tale than is commonly told and it seemed a deceptively simple task to check it out. That simple task led to several years of research. The full story of the search for the history and the people of the legend, and the unravelling of the tangled knot of the actual story of the Lost Dutchman Mine, is told in *Part I* of this work, *The Lost Dutchman Mine of Jacob Waltz, The Golden Dream* published separately (See below).

For our purposes here, however, I am only going to examine the remarks of Higham and Brownie. First, let's look at Higham's remarks. Higham was a newspaper man with a considerable interest in lost mine and treasure stories of Arizona. One of the three manuscripts deposited at the Arizona State Library in Higham's name was a collection of his lost mine stories that had appeared in the newspaper. Higham took a particular interest in the story of The Lost Dutchman Mine and Jacob Waltz, which leads us to one of those great stories that rarely find their way into print.

Higham had written his own version of the Lost Dutchman tale, *The True Story of Jacob Walzer and his Famous Gold Mine* (1945). However, Higham reputedly lost the rights to his book in a legal dispute with a rancher named Barney Barnard. The story is, that while caretaking at the Barnard ranch, Higham ate Barney's prize chickens. When Barnard returned and found his prize stock decimated, he almost strangled Higham. Apparently Barnard sued or threatened to sue Higham for eating his chickens. In settlement of the legal dispute that followed, Higham surrendered the rights of his book to Barnard. One can only hope that those chickens were worth it, as with essentially no significant changes that book went through at least four editions and six printings.

After losing the rights to his Dutchman book, Higham refrained from publishing two other manuscripts, later filing two unpublished manuscripts and his copy of "The Holmes

Manuscript" with the Arizona State Library at the State Capitol in Phoenix. All three manuscripts were promptly locked in the safe, but never catalogued. There they remained uncatalogued and thus "lost" for nearly 20 years. In the early 1970s, they were rediscovered when the vault was cleaned out.

Two things struck me as very curious about the whole Higham matter and *The Manuscript*. If Higham had thought the manuscript worthless, why did he submit it with two of his own works for preservation? Secondly, one had to question Higham's understanding of history as related to the Lost Dutchman Mine since Higham's version of the Dutchman tale included a massed Apache attack on early Phoenix that never happened — facts easily verifiable. It seemed quite likely that Higham's highly romantic stories were just not compatible with Brownie's simpler account and perhaps Higham simply sequestered the manuscript — out of sight, out of mind. In the end, Higham's credibility in assessing the manuscript simply did not hold up, while his efforts to preserve it suggested that regardless of his rhetoric, he believed it had merit.

But, what about Brownie's disclaiming *The Manuscript*? Two possibilities suggested themselves: either Brownie truly did not write it and it was a fake; or, for some reason Brownie wanted to distance himself from the very manuscript he had written. The latter has proved to be the case. Interviews with Brownie's family have revealed that not only did Brownie talk to them of *The Manuscript*, and not only does the family still have Brownie's copies of *The Manuscript*, but they also remember Brownie walking around (pacing?) dictating parts of *The Manuscript* to a typist at their kitchen table. But why would Brownie have created a manuscript only to deny it?

In 1944, Brownie had a crippling accident when a horse rolled on his foot. Brownie believed his lost mine hunting days were done. *The Manuscript* literally grew out of this accident. It seems that Brownie, believing his Dutch Hunting days were finished, had the idea to publish a manuscript about his family's involvement with the Lost Dutchman Mine. His motives were not simply nostalgia. Apparently he thought he might stir-the-waters, so to speak, attracting the attention of someone who might have a missing clue(s) with which he could locate the mine. He would then go into partnership with that person, give out his final clues and they would claim the mine together. (Note: In the fourth paragraph of his Introduction, Brownie speaks of "...taking part in the discovery," not making the discovery.) It was a difficult task. Brownie had to give out enough information to attract the right person, without giving away too much. In the end, the effort failed.

A careful reading of *The Manuscript* reveals two other ancillary motives. Many lost mine seekers, successful or not, when they are no longer able to search actively and physically, like to record their efforts and adventures, and Brownie was certainly one of these. Additionally, it seems that Brownie wanted to set the record straight, as he saw it, regarding his father, for some people claimed that Dick Holmes stole Waltz's gold during Waltz's funeral. Brownie, on the contrary, always claimed that his father was given the gold by Waltz.

Although completed, *The Manuscript* was never published. The reasons for Brownie's failure to publish are not known for certain: Was it because Brownie's foot healed better than he thought it would and he hoped to do more Dutch Hunting himself? Or, was it because *The Manuscript* failed to reveal enough information to attract the right kind of help? Or, did it reveal too much information? Or, . . .

Brownie was often pestered, sometimes endlessly, by people about what he knew of the Lost Dutchman Mine and about *The Manuscript* — Was it his? Which bits were right? What did this mean? What did that mean? And so forth. In the end, Brownie simply seems to have tired of being repeatedly hounded about it, and tried to pretend *The Manuscript* had never been written.

Since *The Manuscript's* rediscovery, excerpts of *The Manuscript* have been periodically published. First, part of one chapter was used as a filler chapter in Robert E. Lee's *The Making of the Motion Picture, The Lost Dutchman Mine* (1975). Subsequently, other authors have taken pieces of the manuscript and combined them with their own ideas and other stories, weaving a tale that Brownie never told.

The Manuscript is presented here in its entirety with family photographs for the first time, telling Brownie's story, not someone else's. Corrections have been limited to spelling and punctuation. Photographs, line art and an index have been added. Additions include a short preface introducing each chapter, endnotes expanding and referencing the subjects of the chapter, and mid-chapter additions or annotations, which are denoted in brackets: [ ]. *The Manuscript* is followed by an Epilogue bringing it up to date, as it has been over 50 years since it was composed, and it has been indexed.

The Holmes Manuscript is presented as Part II of a two volume work on the mine. In *Part I, The Lost Dutchman Mine Jacob Waltz, The Golden Dream.* are the stories of: the people and families involved — the Holmes, Roberts family, and

Peralta families, the Starrar brothers, Julia Thomas and the Petrasch family — the life and times of Jacob Waltz, the origin of the legend, the analysis of the Dutchman gold ore, accounts of frauds, dangers of the mountains, tales of those who have searched for the mine, plus clues and maps related to the Lost Dutchman Mine. In *Part I* you will find not only the story of the Lost Dutchman Mine, but part of a lost or denied history of Phoenix and central Arizona. For your convenience the two books — *Part I, The Golden Dream* and *Part II, The Holmes Manuscript* — are cross referenced

Acknowledgements are due many people and a list of such acknowlegements is presented in *Part I, The Golden Dream*. However, here I should like to especially acknowledge the assistance and support of the Holmes family. They have supplied many photographs for both *The Manuscript* and *Part I*. They have provided details and assistance in many areas over the past few years. I thank them for their help and friendship.

T.E. Glover
Jacksonville, Oregon
April 2000

Once a treasure takes hold of a
man's mind, nothing can dislodge it.

Honoré de Balzac

Brownie's Introduction

Brownie begins by introducing himself, his family, and his involvement in the search for the Lost Dutchman.

So much that is inaccurate has appeared in periodicals throughout the country [about the Lost Dutchman Mine] that I have decided to set matters right by giving a detailed account of the happenings over many years, which provides the documentation for its very existence.

For many years, I have been approached by people seeking information on the location of the mine, as related to my father, Richard [Dick] Holmes, by Jacob Waltz. I will try to do this without disclosing some of the important secrets which have been in the possession of my family for about fifty-five years.

As you follow me on many trips into this remote range of mountains, I will take you many times to a place which, in my opinion, is very near the hidden mine. I will attempt to take and keep the trips in sequence and will relate only what I know to be the truth. I have been asked many times if I know where the mine is. I can only say "NO, but I know many places where it isn't!"

It has been said that the discovery of the mine might seriously affect the gold markets of the world, because of the size of the deposit and the richness of ore. However, I am perfectly willing to take the chance of altering the market if at some time I am fortunate enough to take part in the discovery.

It was in the year 1847 that my grandfather Holmes left his birthplace in Holmes County, Mississippi, and made the long, weary and adventuresome trip into the wild and little known Territory of Arizona. He first located in the frontier town of Fort Yuma, situated on the bank of the mighty Colorado (See: *Part I, The Golden Dream*, Ch. 16).

After spending two years in this community, where the six shooter was the only recognized law enforcement, he moved to the California gold strike of '49. He later returned to Arizona and went first to La Paz, then to Gold Road and in 1863 he moved to old Fort Whipple, thirty miles north of Prescott and there established his residence. My father was born in 1865.

My mother came to Prescott by ox team in 1870. The two met and were married in 1889. Their union brought into one common bond two persons who were from two pioneer families of the Territory.

I have mentioned the foregoing for no other reasons than to show that my people were in a position to relate, from first-hand knowledge, certain events which have occurred in the lives of those people in the early days of the Territory of Arizona, and more particularly to discredit statements that Jacob Waltz, the last man to work the now famous Lost Dutchman Mine, died before my father came to Arizona.

Many such statements ran rampant, particularly when the sensational episode, the finding of Adolph Ruth's skull, took place in the Superstitions. The reported finding of an enormous gold nugget recently caused many people to recall some particular story about the old German and his mine.

In 1931 I acted as a guide to the Arizona Republic's party in the Superstitions when they went to view the Prehistoric City. It was on this trip that we found the skull, later identified at the time as that of Adolph Ruth (See: *Part I, The Golden Dream*, Ch. 18). It naturally got nation-wide publicity, along with Harvey Mott's articles on the nine or ten days we spent in the mountains. Around the campfire at night, I told them of my father being at the bedside of Waltz and a part of the very interesting conversation. This opened up a new field for writers who, to try and give some originality to their articles, continually invented new characters and "Facts" about the mine. Prior to the time, no books had been written about the Lost Dutchman Mine. Harvey Mott made very few notes on anything, except of a scientific nature, and in his articles he spelled Jacob's name as Waltz, instead of Wolz, showing that the Arizona Republic's articles were a background for what was later written about the mine.

In several stories that have appeared about the Superstitions and Jacob Waltz, the writers have tried to discredit my father in one way or another. Let me make this statement: "There was never a map made by the German, Jacob Waltz, showing the location of the mine. Neither did my father ever make or sell a map to anyone, nor did he ever accept financial aid to further his search for the mine. If any map existed, showing the location of the mine, I, or any of the Barkley family (See below) could take you to any direct and designated spot."

I have received many offers of financial backing through the mail and from local acquaintances, but only twice have I accepted a grubstake to further my search for the mine. The time I have profited financially from my knowledge of the Superstitions was from the two hunters, whom I guided on a two-day trip in a search for wild hogs, and the Dons Club paid me to appear at one of their picnics. My name and pictures have appeared in several hundred newspapers in the United States, *The Saturday Evening Post*, three issues of *The World Wide Magazine* in London, newspapers in Sydney, Australia, and South Africa, without my permission.

I have worked during roundups for my friend Tex Barkley, owner of [the] Quarter Circle 'U' Ranch located in the Superstition Mountains. This ranch is owned by my friend Tex Barkley who has the stock grazing permit covering the entire range.

Brownie Holmes
circa 1947

Chapter 1
The Superstition Mountains

The Superstition Mountains are so unusual in their appearance compared with the surrounding flat desert that they almost demand wonder and supposition. The western end of the mountains rises abruptly in sheer cliffs in a series of bastions, not unlike the Cathar castles of southern France.

Brownie begins his narrative by describing the mountains that held him in thrall throughout his life.

These mysterious and awe-inspiring mountains were named Superstition because of the many and varied stories told by the Indians. Strange and weird things were said to have happened within the hidden recesses of this uncanny range.

I, too, have experienced strange and curious happenings within their depth. I shall describe some of them, cause and effect.

The Superstition Mountains have a peculiarity of making a person feel unkindly toward them, even though he may not have heard of the strange happenings which gave the range its unusual and mystifying name.

Towering majestically, like a mighty sentinel of the great Arizona desert, these mountains overlook the broad valleys of the Salt and Gila Rivers. The famous Apache Trail

Superstition Mountain in a late afternoon winter storm.

Highway, over which thousands of people pass each year on their journey between Phoenix and the great Roosevelt Dam, winds its way along the very edge and through the Superstition range.

From a distance they appear as a faint blue haze in the bright Arizona sunlight and then, as one approaches nearer, their immensity becomes more pronounced, and at last one sees them as they actually are, a mighty mass of jumbled rocks, covered with cacti, mesquite and other growths.

"Hell's Back Yard" is a fitting name and description of this strange country, which has at some time in the remote past been subjected to tremendous volcanic upheavals, leaving an area of over one hundred square miles of peculiar vari-colored formation, many of them ghost-like in appearance as evening approaches.

From the hidden valleys of the interior to the gigantic towering peaks, which [rise] nearly five thousand feet into the heavens, these mountains have an appeal because of the very nature of their formation and beauty.

Constant erosion has produced extremely narrow canyons hundreds of feet deep and so narrow in places that one can well imagine himself making a safe jump across the crevice and arriving at a spot only a few feet distant. To arrive at this spot, however, with any degree of safety, requires a full day s journey by animal or on foot through canyons and over mountains.

It is well-nigh impossible for me to describe, with any degree of safety, or rather, accuracy, the rugged interior of the mysterious range, with its abandoned caves and cliff dwellings, nestled high up in the canyon s walls, the remnants of a once-thriving civilization in its valleys, disclosed by ruins and broken pottery and human skeletons. Some of the canyons are so overgrown with flora that they are almost impassable. Yet, for all these drawbacks and the many more which I shall describe later, I love the Superstitions.

As one travels alone within the boundaries of this Devil s Playground , he seems to see the ghost of Peralta, the Spanish Don, whose fabulous wealth is said to have come from a mine somewhere in the interior; then the Jesuit priests of Mexico and their laborers; then he sees the three Mexicans, the soldiers, the old German s nephew, La Barge, Weaver, Pipps, Charlebois and at last the ghost, still fresh in the memory of thousand[s], of Adolph Ruth, the last man of record to pay the supreme penalty of being so bold as to defy the ancient curse of the Superstitions.

[Here Brownie makes the assumption that all the people to whom he is referring are known to the reader. The Peralta family, the "three Mexicans", the (two) soldiers, Waltz's nephew, Pipps and Adolph Ruth are all participants in the legend discussed by Brownie as his story unfolds. Charlebois and LaBarge were early cattlemen in the Superstiton Mountains. For reasons long since vanished, the Jesuits are rumored by many to have sequestered a vast fortune in the Superstiton Mountains, and perhaps this legendary association of the Jesuits and the Superstition Moutnains is that to which Brownie is referring. Brownie's accounts of the Peralta family, the two soldiers and Pipps, Pauline Weaver and Adolph Ruth are all fleshed out in *Part I, The Golden Dream*, Chs. 6, 8, 11 and 18 respectively.]

Dead mesquite tree. One of the treasures of the Superstitions — standing against a dark blue sky, they are majestic.

On the northwestern side of the mountains, a strange freak of nature can be seen. It is a long row of stone figures, cut out by nature during countless years of erosion and which resemble human beings. If a person happened to be alone in the mountains and was of disposition to permit his imagination to get the better of him, the supernatural influence of the country would begin to play its devilish pranks upon him, and if the time happened to be near the close of the day, these figures would seem to come alive and actually move. This apparent mirage would then be accompanied by other figments of his imagination, such as faint cries emanating from the hidden canyons and low moans, not unlike the wailings of a lost soul. These sound[s] and [the] description of their origin will be related further along in the story of my many trips in search of the hidden mine.

Indian tradition has it that these figures represent the petrified bodies of Indians, who long ago sought shelter on the mountain tops, when the earth was besieged by a mighty flood, which filled the valleys below and crept slowly up the sides of the mountains. These Indians were warned by the spirits to remain in absolute silence until the flood waters had completely receded, under penalty of severe punishment. When the crest of the flood had been reached and the waters began to subside, the Indians were unable to hold back their joyous feelings of being saved from the watery grave. The spirits were very displeased at such disregard to their warnings and, as a punishment for this act, turned them to stone, where they are standing today as a warning to those who might talk too much about the legend of the Superstition[s].

One particular portion of Waltz's story which you will read in the main part of my narrative, was verified by Indian accounts concerning a great fight with people of another land, who were slaughtered by the Apaches for their livestock. James M. Bark, a fine old pioneer, now deceased, was for many years the owner of the Quarter Circle 'U' Ranch, which is now owned by my old friend and companion, 'Tex' Barkley, and located on the south side of the Superstition Mountains. Another rancher friend of Bark's, whose name I do not recall, maintained in his employ an old Apache Indian by the name of Jim White [aka: Apache Jack], who not unlike others of his race, was not very communicative about tribal traditions or personal experiences.

Jim White liked and respected his employer and for many years accompanied him on a number of trips over the vast rangeland of the north and after many years of fairly close association with the old man, he gradually let out enough information to convince the old rancher that he knew pretty much about the even-then famous Superstition country.

The rancher told his friend, Bark, about the information he had accumulated and between the two, a gentle sort of conspiracy was formulated. Bark visited his friend's ranch and in the presence of the Indian, inquired about the many types of peculiar game said to roam the interior of the mountains. After considerable discussion, they asked the Indian to guide them there on a hunting expedition to find the midget deer, the wild hog and other animals.

The old Indian consented to accompany them, and with a pack outfit they penetrated deep into the range. After spending a couple of weeks with their guide, establishing a closer bond of comradeship with him, they began to sound him out cautiously regarding the reputed mine. He informed them frankly that he would not talk about it, but he did consent to tell them the story, as an active participant in the bloody battle that wiped out the Peralta expedition.

Many Mexicans came, he informed them, to work a mine far in the interior of the mountains. There were many workers and soldiers with women and children and many mules loaded with supplies. They were driving some cattle and had many fine horses and saddles.

The Mexicans camped at a place with an excellent natural defense in the form of steep canyon walls on either side. The Indians could not hope to make a successful attack as long as the party was secure in their natural fort.

Soon the word spread among the Apaches from far and near that there were many good mules to be eaten; many good horses and saddles to be taken, when the time and place and the advantage in numbers of warriors made it possible to successfully attack the party.

Signal fires and smoke signals told the story. More and more Apaches came and camped here and there in the mountains.

While the Mexicans worked and the weeks and the month[s] passed, the Apache Chiefs took counsel and carefully planned the location of the place where the attack was to be made.

This spot was finally decided to be a little rincon, northwest of the main Superstition mountain, near what is now First Water Ranch. Here all the party could be completely surrounded by the Indians and all avenues of escape would be cut off.

Finally, the Mexicans had taken out enough ore to load their pack mules and started out of the canyon. The Indians arranged themselves at various vantage points to cut them off in front and behind the little rincon with an impassable mountain on either side. A terrific three-day battle ensued and all of the Mexican party was supposed to have been killed. The stock and other equipment was taken by the Indians, and the gold was thrown aside.

This story was in complete agreement in its major details with the story of the fight as related years later to Jacob Waltz by one of the three Mexicans who was working the mine at the time Waltz discovered their camp. This Mexican was identified as the only survivor of the Mexican party after the terrific three-day battle with the Indians.

* * *

[New evidence concerning the Peralta family and supporting the massacre has recently been uncovered. This evidence along with a more detailed account of Apache Jack is detailed in *Part I, The Golden Dream*, Chs. 6 (the Peraltas) and 7 (the massacre and Apache Jack).

While it seems most of the cowboys and cattlemen who have ridden the Superstion Mountains have left few footprints, two of the area's earliest ranchers/cowboys -- John La barge and Charlebois -- are remembered today in La Barge Canyon, the largest canyon in the mountains, and Charlebois Spring and Canyon (🚶). In hot, dry weather, Charlebois Spring is one of the few reliable water sources in the mountains. (See: the United States Geological Survey's (USGS) topographic map, Weaver's Needle.)]

🚶 : Trip 31, Marsh Valley Loop. (Note: Where possible, locations in this work are cross referenced to the *Hiker's Guide to the Superstition Wilderness,* by Carlson and Stewart. Not all locations or clues can be referenced to published trails; those that can be are marked with this format.)

Chapter 2
The Story of Jacob Waltz and His Mine

Brownie describes how his father was first introduced to the story of the Lost Dutchman Mine while his grandfather was out prospecting. On a stormy night, the life of Dick Holmes and his yet-to-be-born son changed forever through the chance revelations of an unexpected visitor.

The "Bloody Basin" country lies approximately seventy miles north of Phoenix. It is yet a very sparsely settled country in spite of the fact that it is excellent range land. It was here that my grandfather and his son, my father, established their horse and cattle ranch about the year 1882. The ranch house was erected on the banks of a creek which was fed from several living springs. The older maps still show this small stream as Holmes Creek.

Although the ranch prospered from the very beginning, there were times when it did not require the close attention of both men and, as a consequence, Grandfather Holmes would spend a considerable portion of his leisure time

Where it started: The remains of the Holmes Cabin on Holmes Creek. All that survives is the stone foundation and flagstone floor. The white rectangle is a watering trough for cattle recently built by local ranchers in the center of the old cabin site.

prospecting, as did a number of ranchers in those days. Sometimes he would be away from home for several weeks and Father would remain at the ranch alone.

In the winter of 1884, Grandfather departed for the hills, to be gone for a period of several weeks. Father was the type of man that enjoyed the companionship of a fellow being, and after spending several days and nights without having seen any sort of humanity, not even a friendly Indian, he was overjoyed one cold winter evening when John Pipps, a prospector, rode up to the ranch house and called a friendly "Hello".

Mr. Pipps was a genial old fellow of about sixty years of age. He carried himself gracefully in spite of his advancing years. It was not the first time that Father had met the old gentleman, for he was in the habit of stopping at the various ranches throughout the country on his frequent journeys to and from the south.

Walnut trees: Less than 50 yards up the creek and on the other side from the old Holmes cabin foundations are the remains of this old walnut grove. (Courtesy: Kraig Roberts)

On this particular evening, the area was visited by one of its regular and sometimes rather severe winter rains. The cold, biting wind whipped unmercifully down the canyon from the north, and the accompanying rain, driving in sheets, seemed to chill one to the bone. Father assisted the old man in getting his outfit sheltered. He then built a roaring fire and, after placing a large tub under a leak in the roof, set about making the visitor comfortable.

They talked about things in general as Father prepared the evening meal. He inquired about people they both knew and wondered what had become of so and so. After the supper, a large sack of black walnuts was brought forth. These he had gathered the previous fall from the many trees in the "Basin" country.

The old man's clothes were dry before long and, as the two were now feeling very comfortable in the cozy cabin, with the wind howling outside and rain whipping itself against the roof in occasional torrents, they resumed their conversation.

Dick Holmes at 13 year.s. Holmes was probably about this age when he first heard the story of Waltz's mine from Pipps, and he would have been only about two years older when he tried to follow Waltz. (Courtesy: Holmes Family)

"Dick," Mr. Pipps exclaimed suddenly as if an inspiration to leave the cabin right then and there had come over him, "what would you say if I told you that I could take you to a place where you could fill that tub with pure gold?" Father gazed at him quizzically for a moment and with a smile replied, "I would say, 'Let's go at once!' "

He had heard sudden bursts of enthusiasm before, coming from the mouths of old prospectors, who had just found the thing they had been searching for these many years. Even his own father had come home on many occasions with a mere handful of pay dirt declaring there was plenty more where that came from. Somehow they did not go back to the place where the rich find had been made, for in the meantime someone had told them of another rich strike in another direction, and so it was with all prospectors.

Pipps continued with his remarkable story, "In the big rough mountains south of the Salt River is a very rich gold mine, Dick. An old German from Phoenix is working it secretly. I have watched the mine for three or four years -- ."

Here Father interrupted with, "Who is this old German, John?"

"His name is Waltz, Jacob Waltz," the old fellow replied, "and he's an ornery old devil too. Why he's killed a dozen men I suppose. The only time I've ever had nerve to go into the diggin's is when I know darn well he's out of the hills. I built a small rock shelter in a cliff, high above the floor of the canyon, where I would be protected from the weather and other dangers. When I was sure that Waltz was not in the mountains, I would dig out some gold and get out of there."

"How much gold do you think is in that mine, John?", Father asked.

"Plenty", the old prospector replied, "That's the richest vein of ore I've ever seen. Why, Dick, you can take out a thousand dollars a day easy."

"How long have you been watching the old Dutchman?"

"For several years now. I haven't been in there for a year though", he remarked.

"Why not?"

The old fellow laughed and remained silent for a moment. "You see, it was like this", he explained. "I thought the old devil was well out of the hills, so I began robbing the

Pipps' Hide Out: Brownie Holmes in front of what he always believed was Pipps' hideout. (Courtesy: Holmes Family)

mine. I had just finished for the day and was going to my hideout on the side of the canyon to rest. I looked down the canyon and — by God, there he was coming back."

The old fellow shuddered as though a sudden chill had gripped him. "You know Dick", he continued, "I'm not a coward but — by God, I don't want another experience like that one. The way that old Dutchman acted when he saw that someone had disturbed his mine — man, it was terrible!"

"How did he act?" Father inquired.

"He began looking for tracks, and knowing if he were able to trail me, he would shoot, I kept hid and when night came I sure got away from there fast."

"Why don't you locate the mine and record it?", Father asked. "Not me," replied old John, "That old Dutchman has killed many a man and he'd kill me in a minute when he found out I located the thing. Nope, I'm getting old, but I've still got my right mind. Dick! By God, I'm afraid of him and I'm not going back until I know damn well he's six feet under the ground."

"Well, I'm not afraid of him," Father declared emphatically. "If you show me where this mine is, I'll locate the damn thing." He was yet to learn of the ruthless and vicious character of Jacob Waltz.

His explosive declaration seemed to startle old John for a moment. He looked Father straight in the eye and said, "Dick, the German is nearly eighty years old and he won't last much longer. When he dies, I'll take you in as a partner and we will locate the thing. I guess I've already said too much." He closed up like a clam.

Try as he would, Father could not get any more information out of John Pipps regarding a certain gold deposit, which was destined to become the most talked-of mine, real or imaginary, in the history of the West: a mine which, during the passing years, has been responsible for the death of at least one man and which sent my father on many disappointing trips into the Superstitions and later myself, his son, on the same quest; always hoping, always believing it was just over the next hill.

Pipps remained with Father for two weeks, during which time he maintained intense silence every time Father brought up the subject of the mine in the Superstitions. He was congenial enough as he helped with the ranch chores and discussed many subjects pertaining to things in general, but so far as the old German's mine was concerned, he was through.

After leaving the ranch, he made his way into the Mazatzal Mountains, a wild remote range, far to the east of the basin. He spent the balance of the winter alone, prospecting here and there, but never discovering anything except an occasional piece of rock that disclosed some colors when panned. As spring came on, he joined a bunch of cattlemen and assisted them in their spring round-up.

The cowboys established their camp one evening on the bank of a creek and Pipps, in his usual industrious manner, set about digging the sand from the side of the deep water hole. He had no more than started when the sand from the sides of the hole began to cave in upon him. He fought frantically to disengage himself from the firm clutches of the wet sand, which had now covered him up to the waist. His cries for help attracted the attention of the cowboys and they immediately rushed to his rescue, but before they could reach the unfortunate man another avalanche of sand fell on him and covered him to the shoulders.

He was quick to realize the terrible predicament of being almost completely submerged in the deadly wet sand and, becoming almost to the point of hysterics, offered a lavish reward for his immediate rescue. He told the startled cowboys, in a gasping breath, that he would take them to the richest gold mine in the world — he would show them more gold than they all could carry out — if they would only hurry and get him out of the hole, which was slowly but surely claiming him as a victim. The cowboys were momentarily stunned at the thought of the impending fate of the aged prospector and, without even considering the promised reward, rushed to their saddles for ropes with which to pull him from the hole. Before they could get back to render him aid, another cave-in occurred and he was smothered in several feet of sand.

There in the bottom of a water hole, high in the Mazatzals, with the promise of gold on his lips, they left his body, buried in a living grave, dug by nature and filled by the hand of Providence; and with him was buried a secret known to only one living man at the time — old Jacob Waltz.

The cowboys conducted a thorough search through his personal effects for some clue which might disclose the whereabouts of relatives. They were unsuccessful in their quest for the information as there was not a scrap of paper in his belongings. The search revealed among other things, however, two one-pound baking powder cans. They were filled with gold!

Father made frequent trips to Phoenix during the five years following the death of Pipps. It was during this five-year period that he met my mother and they were married in 1889.

Shortly after his marriage, Father disposed of his "Bloody Basin" holding and moved to this thriving little western town, which even then was well on its way to becoming one of the most progressive and metropolitan cities in the inland southwest.

About the year 1888, he met an old German who was called 'Old Snow' by the people of Phoenix. They struck up a friendship and after having made several inquiries regarding the old man, [Father] was surprised to learn that he was Jacob Waltz!

Father was very discreet in his conversations with the aged German, in that he never mentioned anything whatever regarding the mine in the Superstitions. He felt that the old man would take his secret to the grave and, inasmuch as the sole possessor of the fabulous deposit of ore was getting very old, he began to wait for an opportunity to learn some vital bit of information.

Old Jake would disappear at the first sign of winter, and the rumors were generally circulated around the town that he was in the possession of a gold mine from which he got the gold necessary to purchase several choice pieces of property in and adjoining Phoenix. [Waltz's winter sojourns were well known in early Phoenix. See *Part I, The Golden Dream*, Ch. 12]

He always had plenty of money on his person and those of a later day have said that he often drank heavily, gambled freely, and boasted of his rich mine in the Superstitions. According to statements made by my father and others who were fairly close to him, he was not of that disposition. On the contrary, he was very secretive and took extraordinary pains to guard his every movement for weeks before disappearing into the mountains.

Small quantities of foodstuffs were purchased over a period of weeks before the old man departed for the mountains. This he accumulated with the secure feeling that he was unnoticed by the curious inhabitants. His one failing, which was the cause of a sometimes close watch being kept on him, was that he became somewhat over-enthusiastic just before departure and made many purchases in the course of two or three days.

After having secured a sufficient amount to last him throughout the winter months, he would make his departure under cover of darkness and would often approach the mountains by a circuitous route. Sometimes he would camp for several days by the side of some remote water hole and watch diligently for a possible follower.

My father began to notice the old man's activities and

he became extremely interested in the various rumors circulating around the community regarding his exploits. His memory carried him back to that cold winter night when the old prospector, Pipps, told him he could take him to a place where he could fill the tub with pure gold. He also recalled the tragic incident of the water hole cave-in when Pipps became frantic and offered to take the cowboys to the richest gold mine in the world.

Came a day when a friend, whose name was McFall, met Father on the street. They talked for a few moments about things in general and nothing in particular when Jacob Waltz approached them with several bundles in his arms. The old man was the first to speak.

"How do you do, gentlemen?" he spoke in his slow broken drawl.

"Fine," Father and McFall replied, "and how are you these days?" Father asked.

The old man mumbled something about the weather and continued on his way. He showed every indication of having been an exceptional man in his early days. His broad shoulders were bowed somewhat and his gait was slower perhaps than that of an average man, but he carried his nearly eighty years remarkably well.

As he turned a corner on the way to his home, McFall turned to Father and said, "Dick, let's you and I follow Waltz into the mountains. He's not going to last much longer and we might as well try to locate his mine before it's too late."

"It's a go, Mac," Father said.

"We'll have to keep a continual watch on him night and day from now on," McFall answered. "If you'll watch him during the day I'll keep him covered at night."

"Fine!" came back Father, "I'll start in right away and watch him until dark, then you carry on where I left off."

McFall looked disconsolate for a moment and then said, "Dick, I'm not much of a hand at trailing and I think I might do more harm than good in following the old man. It would be fatal if we were caught in the act."

To this Father agreed. "I've had considerable experience, Mac, and if you'll just keep me posted as to his time of leaving I'll do the rest."

Father watched the old man closely from morning till night.

The two men seldom spoke now as McFall would take up his duties when darkness came on. They would merely pass each other on the street and nod their heads in recognition.

It was almost two o'clock one morning when father was awakened by a knock on the door of his home.

"Dick, wake up! It's me, McFall."

Father dressed hurriedly and opened the door to let him inside. He could see by the excitement on his friend's face that the Dutchman had departed.

"What's up?" he inquired.

"Jake Waltz has left," McFall answered. "Gone about an hour."

"Which direction did he go?"

"Southeast, toward the Salt River."

"Did you trail him?"

"Yes, about two miles. Once he stopped and I thought he'd seen me. For God's sake Dick, be careful."

"Just where did you leave him?"

"Right where the old wagon road forks. He took the one to the left."

"Good", Father replied, "I'll have pretty good sailing at the start anyway."

Father grabbed up some food: biscuits, bacon and some jerky; hastily placing this in a sack, he mounted his horse and departed into the night. He didn't travel along the road, but kept more out in the desert, the better to conceal the sound of the horse's hoof beats.

Arriving at the spot where McFall had last observed the old German, he dismounted and concealed the horse in a thicket. He then started off on foot and after trailing the pack mules for a couple of miles, returned to his mount and rode to that spot; repeating the maneuver of the "ride and tie" method of trailing. That is, by riding a way and then walking on foot to the point about two miles distant where the horse is to be ridden again. This method of trailing considerably lessens the danger of detection.

The first night after his departure from Phoenix, Waltz camped on the Verde River about two miles north of the place where Granite Reef diversion dam now stands. He unpacked his mules and hobbled them and then set about preparing the evening meal.

As night came on, the winter cold became more intense. Father could not enjoy the comforting warmth of a campfire and he envied the old German whose fire was blazing and crackling merrily. He had crept up to within a hundred yards of Waltz's camp, keeping the old man under his keen observation until his arms and legs became numb with cold. He then retreated to his makeshift camp and made

Mormon Flat, in the foreground, where Waltz crossed the Salt River. Due to the high cliffs along the river, this was one of the few places above Phoenix where you could cross the river. (Courtesy: Superstition Mountain Historical Society)

preparations to spend a long, hard, uncomfortable night without having had the satisfaction of a hot evening meal.

Jake Waltz was up early the following morning. Father had spent a very uncomfortable night and was wide awake when, shortly before dawn, he observed a long column of smoke climbing heavenward, which signified the old German was preparing his morning meal.

He waited for nearly an hour after having first observed the smoke before he ventured forth to note the direction the old man was taking in crossing the river. The brush was thick along the river and he had very little difficulty in ascertaining the whereabouts of Waltz without being seen. The trail crossed the river at an angle and Father could see the faint outlines of the pursued man and his mules, just as they were disappearing in an easterly direction from the previous night's camp.

Father munched his hard biscuits and jerky as he traveled afoot and on horseback through the thick cottonwood and mesquite of the Verde River bottom. On two or three occasions, he came within sighting distance of the German and always, when this occurred, he would hold back to allow Waltz to gain distance.

The trail led to Cottonwood Wash and as evening approached, Waltz made his camp at Agua Escondido (Spanish for Hidden Water), so named because the spring is hidden from view. It was a beautiful spot, with the cottonwoods and sycamore trees lined up on either side of the small outlet from the spring; an ideal place for a campground, even in those days of long ago.

Once again an uncomfortable night was spent without a campfire. However, like most of the hardy pioneers of those days, he took the distressing situation as a mere matter of a day's work and thought lightly of it.

Early the next morning, the pursuer and the pursued wound their way slowly toward [the] Salt River, which was crossed a short distance below the point where Mormon Flat Dam now stands, thence up Tortilla Creek to Tortilla Spring, where Waltz made preparation for his third night's camp. This point is now less than fifty yards east of the crossing of Tortilla Creek by the famous Apache Trail Highway. They were now in the northern part of the Superstition Mountains.

Father used every precaution to prevent possible detection after crossing the Salt River. He went on foot to a spot near the spring and then crawled on his hands and knees through the dense brush and cactus for several hundred yards. Arriving at a point about a hundred yards from the German's

Tortilla Spring, where Jacob Waltz spotted Dick Holmes trailing him. Holmes was coming up canyon (from the left) and Waltz was sitting on a rock "shooting his back trail" — that is, checking to see if he was being followed.

camp, he observed a startling spectacle, one which would make many a brave man experience cold chills up and down his spinal column. There sitting on a pile of rocks, with his rifle in his hands, ready for action, and looking back the way he had come, was Jacob Waltz!

There was a ferocious scowl on his face which plainly disclosed the fact that there was murder on his heart; and Father, realizing the jig was up so to say, backed cautiously out of the brush.

There was no sense or reason in pursuing Waltz any further, now that he was distinctly on the defensive. To continue following the old man would mean one of them would not come back. Father would not think of killing the old man. In the first place, Waltz had never injured him in any way. As a matter of fact, he felt more kindly toward Father than he did toward most of his acquaintances — and also if it became necessary to fight it out, as they would be sure to do further along the trail, and if the old man should be killed — right then and there would end the secret of the mine.

Father was thinking of all this as he returned to his

horse about a mile or so distant from the spring camp. It was possible, he thought, that the German had discovered that he was being followed, but it was not probable that he had learned the identity of who was trailing him. Later Father found out that he had been wrong in this assumption.

The following spring, Waltz returned to Phoenix and, meeting Father on the street one day, said in his slow accented drawl, "Dick, I have something to say to you."

"Surely, Mr. Waltz," he returned guiltily.

There was never the least inkling in the old man's speech or actions that would indicate any ill feeling on his part and Father was very much surprised when he said, "You tried to follow me into the mountains, but don't ever do it again, because the next time I will surely kill you. I like you, Dick, and I don't want to harm you, but remember what I've just said."

With that, the old man continued on his way, leaving Father there with the feeling that he was about the lowest creature in the world. He would much rather have had old Jake come to him storming and cursing, for then he could have put up an argument and would have had the satisfaction of having held his own, but to be talked to like a bad little boy, who had done some mischievous trick of not much consequence, hurt his pride considerably. He was a very capable tracker and he knew it. Constant work on the range had taught him all this. He realized the old German was unusually shrewd in detecting the fact that he was being watched and trailed by an expert in the game.

Needless to say, Father did not follow him on subsequent trips. Had he fully realized what a ruthless killer the

Dutchman was, it is very doubtful that he would have trailed him in the first place.

Waltz made his last trip into the mine in the winter of 1891-92. History will tell you that it was one of the worst storm-ridden winters ever experienced in the southern part of Arizona. He was working his mine when the storm broke and, catching a severe cold, he packed part of the ore and came to Phoenix. By the time he arrived here, he was a very sick old man. He was put to bed and was attended by an old Cuban Negress [Julia Thomas].

Despite her efforts, he gradually became worse and pneumonia developed. Realizing that his time on earth was growing short, he called her to his bedside and told her that he was going to die and that he wished to give directions to his mine. Mrs. Thomas became very excited and rushed from the house to obtain help. My father and a friend named Gideon Roberts, a much older man than my father, were the first people she met. She told them of the old German's condition. Father sent her for a doctor, and he and Roberts hastened to the bedside of the stricken man.

When they arrived at Waltz's bedside, he told them to sit down and addressing my father, said. "Dick, my time is short, and I want to give you the directions to my mine."

(Please remember that Waltz spoke broken English and he was not familiar with the entire Superstition range. He had no reason to explore them; he already had "pay-dirt". Even today there are many unnamed canyons and hills within the range and, at the time of his death, there were less than half a dozen named places. Therefore, his directions were limited to the government trail and marker.)

When Mrs. Thomas returned with the doctor, Waltz had passed away, but not before he had told my father and Roberts one of the most fantastic stories ever told about a hidden gold deposit in the history of the West.

* * *

[Brownie's contention that Waltz liked Dick Holmes and was apparently willing to "cut him some slack" and overlook what had, after all, been an attempt by Holmes to steal Waltz's mine is, in the least, interesting. Why would Waltz be so generous? When Holmes tried to follow Waltz, Holmes was evidently only about 14 or 15 years old (see below). Waltz, at 70 plus years of age, could have just been acknowledging the exuberance and vagaries of youth, and that is why Waltz talked to Holmes "...like a bad little boy." Consider how being dressed down like a child publically would crush the ego of teenage boy!

On his deathbed in 1891, Waltz told Holmes that for the past seven years he (Waltz) had been living off of the gold he brought out with him on his last trip from the mine in 1884. Hence, Dick Holmes had to follow no later than 1884. My best guess is that Holmes probably followed Waltz some time around 1880, and since Dick Holmes was born in 1865, he would only have been in his mid-teens.

Although the manuscript indicates that Waltz worked the mine up until the winter of 1892 and his father tried to follow Waltz only a few years before, this is clearly not the case. Waltz could not have worked the mine the winter of '92 as he had died the previous October (1891). Nor is it even likely he had made his usual winter trip the year before in 1890/'91; if he had he would not have been in Phoenix in February of 1891 to be trapped by the flood.

That the chronology in the manuscript is off there is no doubt. It is known that Brownie sought the help of a friend in preparing The Manuscript, and that that friend had a strong sense of melodrama. After a while, Brownie realized that his friend's free help was worth what it cost; and he had to spend hours re-redoing The Manuscript where his friend had ""helped." My guess is that the timeline of Waltz working the mine in 1892 and Dick Holmes not following Waltz until the late 1880s may well be a bit of the residual from his friend's attempt to write a more melodramatic story by stretching Waltz's working at the mine almost up until the time of Waltz's death -- which is, after all, a more dramatic story than Waltz not having worked the mine for several years before his death and spending his last years in quiet seclusion. (See *Part I, The Golden Dream*, Chs. 13 and 16.)

Although Brownie gives only passing reference to Julia Thomas, Julia is a principal figure in the Waltz story, and a rather remarkable woman of early Phoenix. She was a successful business woman, leader of the first searches for the Lost Dutchman Mine and a founder of the New Israelites religious movement. References to her appear frequently in the newspapers of her day. Her story is told in *Part I, The Golden Dream*, Chs. 4, 13 and 15]

Chapter 3
The True Story of the Lost Dutchman Mine of the Superstitions as Told to My Father, Dick Holmes, By Jacob Waltz on His Deathbed

The old man was very sick and frequently, during his conversation, was compelled to rest for a moment, but in spite of his very weakened condition, he related the following:

"Dick, I know that you've wanted to locate the mine. You tried to follow me once and I might have killed you. I'm glad I didn't because I like you, there is no one that I'd rather see find it than you."

"Years ago I was working in the Vulture near

Three Apache watch Waltz start to make coffee

Wickenburg. I had made a little stake and I wanted to make a trip to Picket Post. (Picket Post is located near where Superior is now.) After I got there I didn't want to stay, so I started back to the Vulture by way of Fort McDowell over the Government Trail. This trail, as you know, is used by the soldiers stationed at McDowell. This was in 1877. I had three mules with me and a pretty fair camping outfit that I had bought a few months before."

"The first night out of Picket Post — I guess I was about fifteen miles from there — I made my camp. Next morning about dawn, just as I was about to start a fire to make some coffee, three Apaches came running toward my camp. I knew what the ornery devils were up to, so I grabbed my rifle and let them have it. We played hide-and-seek among the rocks for quite a spell and I knew they'd get me sooner or later, so I slipped out of the arroyo and got away from them. They looked for me for a while, I guess, and then made off with my outfit. That's all they wanted anyway."

"Well, there I was afoot with no water and no breakfast. I had gotten off the trail considerably and was pretty much lost. I knew the general direction to the Fort, but it was a long way off. I had to keep a close watch for other Indians because that whole country was full of them and they were pretty damn mean, too."

"I didn't go very far that first day, because the country was rough and in some places I'd have to walk a half a mile or more to get around a rough canyon. That night I found a cave in the side of a hill. It was full of bats and every once in a while I heard a noise a lot like a rattler. But I'd rather sleep with the bats and the rattlers than to take a chance on them damned Apaches."

"Early next morning I started out again. I hadn't had any grub since the Indians ran me out of my camp and my feet were getting sore. I had just crossed a wash and got up on the other side when I saw human footprints, so I followed them. They looked like they were fairly fresh tracks and they were pointed in the general direction of Fort McDowell."

(Here the old man stopped and rested. He felt his sides and emitted a suppressed groan as if a sudden severe pain had pierced his congested lungs. After a short interval he continued.)

"I had only followed the trail for about a half a mile when I came upon a camp. I could tell by the size of the outfit that there was more than one person belonging to it. There wasn't a soul there, but there was grub and water. It

Vulture Mine. One of Arizona's great gold mines, it fostered its own city at the mine, Vulture City, where Waltz would have stayed. The remains of Vulture City today.

wasn't Indian grub either and after I tasted some of it I knew it wasn't white man's grub. Anyway, I ate plenty. I didn't drink much water though, just took a sip now and then until I quenched my thirst."

"It was just getting about sundown when I woke up. I heard some voices and I jumped up and grabbed my rifle. Pretty soon three Mexicans came into camp. They looked friendly enough but they acted surprised to find me there. One of them, who spoke fairly good English, asked me what I was doing there and I told him about the affair with the Apaches. They all laughed and said I could stay there with them until they were ready to go to Fort McDowell, which would be in a few days, or if I wanted to go ahead they'd fix me some grub. I was just about in the notion of taking the grub and going on my way when I saw one of them unload a sack of ore from a burro."

"I asked them if they would mind letting me see the

ore. They were simple Mexicans and not at all suspicious, so they let me take a look. Dick, that sack was full of gold ore. I asked them where the mine was located and they said I could go with them the next morning. I was still dead-tired when I went to bed, but I couldn't sleep a wink thinking of that gold."

"Next morning I went with them to a place about a quarter of a mile up the canyon from their camp. They showed me the shaft and asked me if I'd care to go down and take a look. The hole was only about four feet across and about twelve feet deep. I walked down the ladder and when I got to the bottom —"

(Here the old man paused again. It seemed to Father that he was reliving the scenes of that exciting day. His feeble old hands shook as he continued:)

"There was gold, Dick—wire gold—pieces of gold in the rocks as large as peas. Several pounds of ore was lying on the bottom of the shaft and I could see where they had been following the vein for just a few feet. It was about this thick"—(Here he indicated the approximate thickness of the vein of rich ore with his hands. He held them about two feet apart.)

"As the Mexicans worked, they talked freely. They told me they had only been working the mine for a few months and that it had been originally operated by one Pedro Peralta, a Spanish Don, who left Spain many years ago and took up his residence near Mexico City. Don Peralta made only three trips from Mexico to the mine. His party would consist of peon laborers, soldiers as escorts, their women and children and many animals including cattle and sheep."

(Here Father interrupted the old man by admonishing him that he should rest for a moment. He wasn't particularly interested in the activities of Peralta and he realized the old fellow was sinking fast. What he wanted was the secret of the mine before it was too late.)

The old man waved aside his protestation and continued his story:

"On these trips every precaution was taken to insure protection from the Apaches, who were always hostile towards Mexicans; resenting their coming for gold and always attempting to raid the party to capture the animals and whatever else they could get. As a means of protection, the Mexicans would herd their stock by day and place them in corrals at night under heavy guard."

"About 1848, after a winter of successful mining, the Peralta Party broke camp and [Peralta] started his party on the long trip to Mexico. They were attacked by a large band of Apaches who were not as well equipped as the Mexicans, yet their number was greater and the Peralta party was handicapped because they had to defend their stock, and women, and children."

"Peralta tried to fight his way into the open country where he would be in a better position to defend the party. The Apaches guessed he would do this and were prepared for it. They tried to make their way to the west and then over the south slope of the mountain where they were completely surrounded by the Indians. All were massacred with the exception of a twelve-year-old boy, who had managed to make his escape during the battle. The lad found his way to a little Mexican settlement on the Gila River. The village people helped the boy make his way back to his people in Mexico. The lad never forgot the rich mine and in 1877 he returned with his two cousins and began working the property. The Mexican identified himself as the twelve-year-old boy who was the only survivor of the three-day battle between the Apaches and Mexicans. The other two Mexicans were the cousins who had come to help with the mine."

"They [the Mexicans] explained to me that the early Spaniards had put rocks in the forks of trees leading from the desert to the mine. Some of these rocks are so imbedded now that they can only be moved by chopping them out."

"They took me into their confidence as though I was one of them. Their simple stories and absolute frankness, together with their extreme generosity, impressed me not in the least. I had the gold lust in my blood, Dick. I made up my mind to get sole possession of that mine at any cost. In a few days I saw my chance."

"I watched them closely. One of them went up to the mine to put some tools away as they were ready to break camp and go to Fort McDowell. I stayed at camp with the other two and made out like I was assisting them in packing their outfit. When I finally saw my chance — they both had their backs to me — I grabbed up my gun and shot them both. I then covered the bodies and waited for the third man to return to camp."

"I didn't have to wait long. It seemed only a few minutes until he came down the wash. I waited until he was fairly close to me before I shot him." (A pause and a noise coming from his throat that sounded very much like a whimper. He was extremely nervous now.)

"I buried the bodies close to their camp, then took their outfit and went on in to Fort McDowell. I stayed there for two days and then came down to Phoenix."

"I made another trip to the mine some time later. I worked it for several weeks and took out several thousand dollars worth of gold."

"Before I made plans for the next trip, I wrote to my sister in Germany and told her to send her son over to help me. I sent her several thousand dollars and told her there was lots more to be had. Soon after my nephew arrived in Phoenix, I took him out to the mine. He hadn't been there long until he insisted on me recording the property and taking out all the gold at one time, rather than make a secret trip every year."

"I was not a citizen of the United States nor had I declared my intentions to become one. For that reason I couldn't locate and record the mine according to law and even if I had been a citizen, I was afraid there might have been other Mexicans that knew of the mine and that they would come to investigate."

"I knew from the start that there would be trouble between my nephew and me, because he was very quarrelsome and pretty much of the type to tell other people what to do. It got so bad we were quarreling all the time and two or three times I thought he would strike me. Somehow I got to hating him as if he were a bitter enemy instead of my nephew. We left the mine one morning and started for Phoenix by way of Fort McDowell and that evening we camped at Agua Escondido. We had a terrible argument that night, the worst we'd ever had. The next morning when he got up he said, 'I'm

Hidden Water Spring. Where Waltz killed his nephew. Dick Holmes later located the grave and the site became well known to the local rancher, Johnny Jones. Jones became a confirmed Dutch Hinter who liked to spin yarns to greenhorns about the mine. Arrow indicates the approximate site of the grave. (Source: Coolidge, *Cowboys of Arizona*)

going to record the mine myself whether you want to or not'." (The nephew did not know the law regarding aliens filing mining claims.)

(The dying Old German looked up at Father when he got to this point in his narrative. Tears were gathering in his fast dimming eyes and he said feebly);

"I shot him in the forehead, between the eyes. I then took a piece of chain and putting it around his neck, dragged him under a shelving of rock and where the dirt was soft I dug—a—grave—and—buried—him."

(The old man was softly crying now. It was evident that the slaying of his nephew unnerved him far more than the other murders he had committed.)

"I returned to Phoenix and when people asked me where my nephew was, I told them he'd gone back to Germany. I took care of my sister though, Dick. I guess I sent her over two hundred thousand dollars altogether. When you locate the mine I want you to give her a portion of it. Will you?" Father nodded his head in acquiescence.

(After the confession of this, his fourth murder, the old man remained silent for several minutes, with his eyes closed; after which he rallied and indicated that there were more killings of which he must confess before he could give the directions to the mine. Permit me to digress from his story for a moment to give a description which I feel should precede his further confession.

(Fort McDowell was one of the main United States Cavalry posts in the Southwest. It was located near the Verde River about 35 miles northeast of Phoenix. There were several other posts in the Territory at that time, one of them being at Picket Post and another at San Carlos. Part of the duties of the soldiers was to escort pack trains through the mountains as well as carry the mail between the various posts. The mail carriers on one route consisted of two men, who made the trip from Fort McDowell to Picket Post once a week. After a series of trips back and forth, the two soldiers began to appear in Silver King [Pinal] with a considerable quantity of gold in their possession. They boasted of having found a place where there was plenty more.

(There came a time when they did not arrive at their destination on schedule. Investigation was made as to their whereabouts and their bodies were found a few days later within a mile to the east of Weavers Needle in the Superstitions. Circumstances surrounding their condition indicated that they had been murdered and horribly mutilated by the Apaches.)

"I had been gone from the mine for several months,

Dick, and on the next trip took with me some lumber to build a rocker as I was wasting some of the finer ore. I had to abandon some of this lumber in a rough brushy canyon, and upon returning to the mine, I found that it had been disturbed. That gold had done something to me. Somehow I felt that I was the sole owner of all the Superstition country. It made me mad to see anyone going in the direction of these mountains. I made up my mind to hide, because I knew that whoever had disturbed the mine would return. I waited and watched for two days, and then two soldiers came up to the mine and began working it. They worked only a few hours and then got on their horses and rode away. I followed and killed them. I made it look like the Indians had done it."

"After I killed the soldiers, I went back to the mine and took out some more ore and then came back to Phoenix. When I returned the next winter, I saw where someone else had been there. I knew then that I was going to have to do something if I intended to keep the mine, so I started concealing it so no one could find it."

"I enlarged the shaft about 2 and 1/2 ft. all around and left a ledge about 6 ft. below the surface. I then went up on the ledge. I worked all winter sawing timbers the right length to fit on the ledge. I worked all winter sawing, dragging and placing those timbers. They're in the shaft now, crisscrossing to a depth of six feet. I left about two feet near the top so I could fill it in with dirt and rocks. No one will ever find it unless he finds my rock house down in a brushy canyon. It's almost impossible to get there unless you know how. But when you do get there and follow my simple directions, you can go straight to the mine. The hideout is well covered with a natural growth of brush and trees and is impossible to see until you are almost upon it."

"Well, Dick I hid some of my belongings in the hide-out, took what gold I had dug up and started out to Phoenix. About 3 miles from camp, I met a prospector with two burros loaded with camping equipment—" (Here the old man hesitated again and maintained his customary silence for a moment, his eyes closed)—and I shot him without giving him a chance to explain where he was going and what his business was in the mountains. I drove his burros quite some distance from where I killed him. I unloaded the equipment and set fire to it, and then drove the burros away."

(Waltz felt that in disposing of the aged prospector he had eliminated the last man who knew of the mine's location or existence. He did not realize at that time [that] there was a man that very moment in Silver King [i.e., Pinal] whose name was Deering and who was gathering supplies to make a return trip to the mine.)

"The prospector's death made seven men that I've killed - - Dick - - seven human lives over — gold — and look at me now."

"My last few trips were made by way of the Salt River, traveling up to Monroy's ranch and taking one of the old Government trails which crossed the river at that point and then following it into the Superstitions."

"I worked the mine for two weeks this winter before the rains set in. It was very cold and somehow the weather seemed to affect me more. One morning I had a chill. I knew there was something wrong and somehow I couldn't help but believe that I was on my last trip to the mine. I hid most of

The Hill Letters

On file at the Arizona Historical Society in Tucson are a series of letters from an Edwin Hill to a Mrs. Kitt. Hill was a close personal friend of Monroy and knew his ranch well. His full name was Francisco Monroy, a native of Mexico born in 1840; died December 5, 1935. The letter dated November 1, 1940 contains the following two excerpts, apparently supplied in response to queries by Mrs. Kitt:

> "You may remember I told you the Dutchman crossed the Salt River at the Monroy ford. Further, that he came there on [the] opposite side of the river, with his burros, to wash, that he had a place for his bed about a day's journey back from the river?" . . . and . . .
>
> "The Dutchman sometimes crossed the Verde [River] at McDowell ford, I was told."

Signed: Edwin B. Hill

the gold I had taken out, and brought the rest with me. There's not much there, but after I'm gone you can have it. It's in a box under my bed—take it—I know there's enough to grubstake you for a while anyway."

"Dick! Dick!" the old man gasped, "I must tell you -- " He breathed laboriously and there appeared a peculiar pallor on his cheeks, as if his life blood was quickly drawn away from his body. "Go to First Water," he whispered, "then to Second Water, then take the old Government Trail to San Carlos. Where—the—trail—turns—south—you—will—see—over the point of a ridge, a rock, standing in—the brush—and it looks like a man. When I would first see it—I thought it was—a man so one day—I rode over to it and—with my pick—carved the eyes, nose—and mouth on it. I said, 'You look like a man, so I'll make a man out of you.' This is where—I always left the trail—."

Government Well, as it was in Waltz's day. Dick and Brownie always considered this to be what Waltz meant when Waltz said: "Go to first water." (Courtesy: Salt River Project)

His strength failed him and once again he lay with eyes closed. Father thought he was dead because the peculiar pallor of his face seemed to become more noticeable, and then he rallied with renewed strength and continued:

Go to the left of the trail. Follow up this long ridge and you will come to a saddle. In this saddle is a round Indian ruin of rocks. Go through the saddle and up on a long ridge and when you get on the highest point of the ridge you can look north and the Four Peaks are lined up to look like one peak. In the other direction you will see a high needle (pointed rock). In the canyon, under you, is my hidden camp. You can t get down there because it s too steep, but go to the mouth of the canyon and then back. You can find the rock house with very little trouble. You won t be able to see it until you are right on it. After finding the camp, then come back out of the canyon and (Here he gave directions to the mine.) "You will never be able to find the mine until you first find the rock house as the shaft is completely hidden. A prospector will never be able to find it because there is no ledge in view. In the mine you will find about $75,000 in gold, already dug out. Be careful in opening the shaft as I have set a trap that will kill any person who accidentally finds the mine."

Here father interrupted with, "Waltz, did you dig all of the gold out or is there some left?"

"There is enough gold there to make twenty men millionaires," Waltz replied, and holding his hands up about two feet apart, continued, "The ore is this wide and runs the mountain about four hundred feet, where it drops out in the bottom of a wash. I dug this cropping away and erased all signs of my digging."

Here the old man stopped and rested for a while. He suddenly appeared to become weak again. When he resumed his narrative, he talked in a whisper and the whispers became more faint as he went on. At times he would gasp for breath and clutch his breast weakly. "Follow my directions and you will have no trouble in finding—the mine. First, find the rock with the face on it—then—the rock—house, you—have to—do—that. Don't forget—to take—the candle—box of ore under—my—bed. When you find—the mine—send my sister—some money and—"

He closed his eyes, taking a deep breath as if to continue his conversation. There was a slight stiffening of the giant frame; a gurgling sound emanated from somewhere in his throat — and Jacob Waltz, sole owner of the fabulous mine of the Superstitions, was dead!

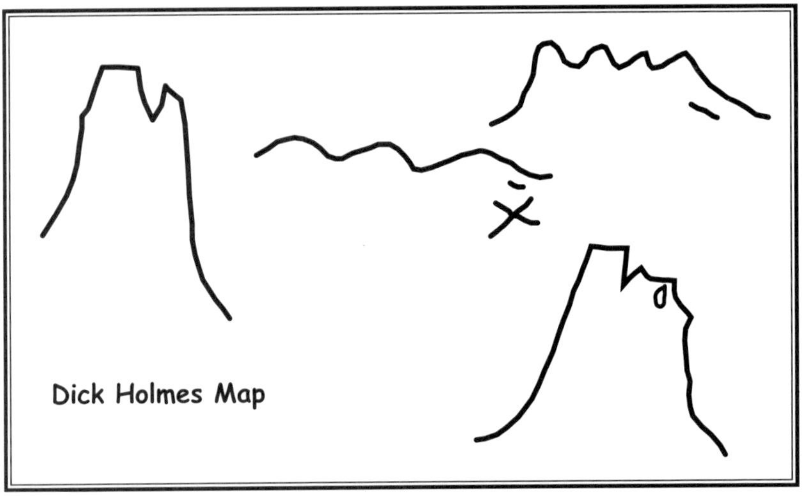

Is this the deathbed map of Jacob Waltz? The question of rather or not Waltz drew a map has been hottley debated. Julia Thomas sold maps she said were copies of a map drawn by Waltz. This map, which appeared decades after Waltz's death with Dick Holmes' name attached to it, bears some resemblence to drawings Rhinehart Petreasch claimed Waltz had done. Brownie Holmes always swore there never was a map, and disowned "The Dick Holmes Map". The question of this map, Julia's and some two dozen other maps is taken up in *Part I, The Golden Dream*, Ch. 22.

Father and Roberts remained by the bedside until the doctor arrived, with the much excited black housekeeper. It was fortunate, so far as listening to the story of the mine was concerned, that the colored woman had some difficulty in locating the physician. However unfortunate it might have been for poor Mr. Waltz, it was very doubtful that anything could have been done anyway to save his life.

The doctor gave the corpse a hasty examination and returned to the bedside of a woman who was soon to present the town the of Phoenix with a new citizen to replace one who was no more.

Father turned from his rather long and thoughtful observance of the corpse of Jacob Waltz to speak to the colored woman, but she had vanished. He then reached under the bed and brought out the candle box of ore, which the Dutchman said he could have. He and Roberts carried the box home and ground the contents up by hand, laying aside some of the specimens of quartz for future study.

The gold content of the high-grade ore was approximately forty-eight hundred dollars, which was a godsend to father at that [time] as he had left the Government service as a packer to be near my mother, who was soon to bring another citizen of Phoenix into the world, no less—George (Brownie) Holmes, and he was at that time running short of funds as a result of his unemployment.

I have in my possession at the present time a beautiful pair of cuff links, each of which contains an elaborate setting of polished quartz, taken from the Lost Dutchman mine. In addition, I also have a tie pin and a priceless ring, together with a number of pieces of the gold quartz.

Waltz was buried amid very simple ceremonies in the cemetery on west Madison Street, in Phoenix. This cemetery is still today in existence, but is in a very sad state of repair. Within its limited boundaries have been buried some of the foremost pioneers of Arizona and it stands today, unkempt and unsightly, instead of being a beautiful shrine, perpetually maintained with flowers and shrubbery as a slight token of reverence and appreciation for those who have helped make Arizona what it is today, and whose bodies are now reposing under unattractive soil, forgotten. In this same cemetery, very close to Waltz's grave are buried my grandmother and grandfather; also an aunt and an uncle are buried there.

Roberts and my Father made a pact that neither would divulge what Waltz had told them and that they would share the mine if either found it. Roberts did not live long, as he was a victim of silicosis, and when he passed away Father alone knew the story of the Lost Dutchman Mine.

* * *

[The question has always been: How accurate are the clues in *The Manuscript*? The best assessment one can presently make about Brownie's clues to the Lost Dutchman Mine are they fall into two general catagories: those he omitted, his secret information, and the clues and information he was willing to make public in hopes some one would find the rock house. Of those clues Brownie included in *The Manuscript*, the best assessment is that some are accurate, others almost accurate and/or accurate but out of order. It is known that the very start of the directions, "Go to First Water" is not what Waltz said. Remember that what Brownie hoped was that someone would find the Rock-That-Looks-Like-A-Man and/or the Rock House. In the end, these were the two major clues, the clues that he thought would locate the mine.

He seems to have wanted people to know about the Rock-That-Looks-Like-A-Man being associated with a Military Trail, and he apparently wanted people to know about a saddle where the Four Peaks looked like one. But, he also seems not to have wanted to give out, to surrender, all the directions in their right seqence. The simple fact is there was little reason for him to disguise the directions since these directions had not done him any good. It is entirely possible, that by this time, Brownie believed that the dying Waltz may have made an error or left something out in his final moments.

Brownie's thinking seems to have been, if he could only get people looking along the old trails for the Rock-That-Looks-Like-A-Man, or get them looking around and below saddles where the Four Peaks looked as one, someone might find the Rock House.

What Brownie did not know was that he was playing against a stacked deck, so to speak. As a cowboy for Tex Barkley, he worked in an environment where almost everyone hunted the Lost Dutchman Mine and Brownie's refusal to tell what he knew frustrated many a cowboy. That frustration turned to anger in some, anger that spawned destruction of possilbe clues.

It was known that one of the things Brownie was looking for was a Rock-That-Looks-Like-A-Man, but just why he wanted it he would not tell. One day a cowboy working for Tex Barkley found a Rock-That-Looks-Like-A-Man in precisely the type of location that Brownie was known to be seeking. Well, Brownie would not tell the cowboy what he knew and the cowboy was not about to tell Brownie anything unless Brownie would divulge some of his secrets. In the end, the cowboy packed an iron pry bar into the mountians and turned over the Rock-That-Looks-Like-A-Man that he had found, tumbling it into a gulley. When the cowboy told the story to Snell (a friend of Roy Bradford, See *Part I, The Golden Dream*, Ch. 17) he reportedly said: "Well, now I know something that Brownie doesn't and he never will!" Was it the right Rock-That-Looks-Like-A-Man? We will never know. For the problem has never been finding one, it has been finding one in the right type of location. What other clues were similarly destroyed? Who knows, but I do know of at least one Rock House that was. Was it the right Rock House? Hopefully it was not.

While Gideon Roberts is mentioned only in passing in *The Manuscript*, his role, and that of the entire Roberts family, is possibly pivotal to Jacob Waltz and Dick Holmes. While the Roberts family was a significant factor in the development of Phoenix, their history had been little studied (See *Part I, The Golden Dream*, Ch. 16).

Concerning the prospector Waltz confessed to killing, almost 50 years after Waltz's death Brownie found the remains of a very old prospector's outfit, which had been burned, deep in the Superstition Mountains, not far from Bluff Springs Mountain.

Besides the directions given in this chapter and clues in different parts of *The Manuscript, Part I, The Golden Dream* has a listing of most of the other known clues and over a dozen treasure and location maps, including an old military map showing the military trails of the Western Superstition Mountains.]

The Nelsons: Jim, his wife, June, and Jim's mother, Lola Madriel. Lola's father worked with Waltz at the Vulture. After introductions, I turned on my tape recorder and listened to Lola and her older sister tell of having grown up in the last remnants of the old west. She told how her father had come up from Sonora in the early 1870s, and worked as a miner and stagecoach guard until the 1880s. He had saved his money and bought a ranch — a ranch still owned by the family. She told how her father had known Waltz. Both men worked at the Vulture Mine, where, according to her father, Waltz was caught highgrading (that is, stealing highgrade gold ore) and fired. That Waltz could have been a highgrader certainly does not stretch credulity, given the more serious crimes he confessed to Dick Holmes and Gideon Roberts. (See Chapter 3.)

I thought I had only come for a brief introductory visit; I was not prepared for the Nelson's hospitality. I ran out of tape as Grandma and her sister talked of the old days — what her father knew of the Wickenberg Massacre and how her younger brother had been killed by Apaches. I stayed for dinner; Jim told stories of the ranch and showed off the family's collection of heirlooms. That afternoon was a wonderful, unexpected pleasure. True westerners, and true western hospitality.

Chapter 4
The First Searches for the Lost Dutchman Mine

The results of Julia Thomas' first hunts for the mine and start of Dick Holmes' searching follow. Holmes realizes that finding the mine is going to be more difficult than he had anticipated. He begins to check Waltz's story.

News of Waltz's death and of the fabulous rich mine he worked, spread rapidly. The title of the 'Lost Dutchman' was given to the lost mine immediately as its fame spread far and wide. Various stories, some of them preposterous, began to circulate as to its origin and location.

[The fame of the mine did spread quickly — within four years of the death of Waltz, there had been at least eleven newspaper articles about the mine, including five feature articles in three states. (See *Part I, The Golden Dream*, Ch. 5)]

Maps of every description appeared frequently in the hands of people who had never before been in the Superstitions. Some of them were so impossible that even the

A QUEER QUEST.

Another "Lost Mine" Being Hunted for by a Woman.

Newspaper announcement of Julia Thomas' first hunt for the mine of Jacob Waltz. The article notes that Julia and two men had been in the Superstition Mountains for two months hunting a lost mine. (*Arizona Daily Gazette*, Aug. 27, 1892)

most tender of tenderfeet scoffed at them, yet they disclosed the fact that some person or persons had already begun to "cash" in on the mystery of the lost mine.

Many people, thinking that the Negress housekeeper of the departed Dutchman was in full possession of his secret, treated her with great respect. This unexpected and unusual attention pleased the woman not a little and she began to manufacture all kinds of stories about the 'Lost Dutchman' and its whereabouts. Her exaggerated stories were responsible for many futile expeditions being undertaken.

On many occasions, she would go so far as to accompany a party on a search, for a consideration, and always on arriving near the range, become confused and feigning illness, insist that she be returned immediately to Phoenix. At last when all her stories were proven to be without foundation, she admitted that my father was the sole possessor of the secret of the "Lost Dutchman of the Superstitions." These statements, of course, turned the attention of the treasure hunters upon him before he had an opportunity to make his first searching trip into the mountains. The Negress later married a religious fanatic whose name I wish to conceal because he has living relatives in the state.

The two [Julia Thomas and Albert Schaffer] began to have visions about the mine and as a result of their 'dreams' (and imaginary consultations with the spirits of the mountains) began to manufacture other fantastic stories from which other variations of the directions to the mine were formed. Soon their minds centered on a religious belief of their own concoction. They turned their home, which was located at the time on the corner of Second Ave. and Jackson St. in Phoenix, into a meeting place where they practiced their religious rites. The woman's husband let his beard and hair grow quite lengthy and then began to appear as the Christ. Among other preposterous statements, they declared they had received word that the Earth was soon to be visited by another great flood, more gigantic than that which appeared at Noah's time. They would climb [on] the roof of their house, declaring themselves safe as that particular residence had been blessed by the Almighty and was to become the Second Ark.

Their antics and outlandish maneuvers attracted many people and after three days of raving and shouting from the house top, the officers moved in upon them and they were compelled to postpone their activities.

During the next fifteen years, the two had acquired a following of a number of men, women and children and, not unlike a number of similar cults, made known their desire to be left alone where they could worship, receive their signs and make their further prophecies without being held up to the ridicule of the unbeliever. Eventually they drifted up to the head of San Domingo Wash, in the Wickenburg District, where they carried on their fanatical rites for several years

The type of Stage Brownie Holmes drove: In the early years of this century, what we term large touring cars were operated as "Stages" over many of the dirt roads in the Arizona Territory. It was one of these that Brownie drove. This illustration is taken from a Southern Pacific poster advertising their services and the "Stages" that connected to their lines. (Courtesy: Doug Stewart)

and where the man finally died, believing to the last that he was the true Son of God.

The two-hundred-pound Negress continued the unfinished 'work' for several more years and at her death was discovered to possess a necklace of solid gold nuggets, valued at nearly one thousand dollars, which was supposed to have been given to her by Jacob Waltz. It was said that the necklace was buried with her body by the members of her cult. It was rumored that the organization disbanded some years later.

Almost immediately after Waltz's death, people began to pour into the Superstitions from all directions. Father felt that inasmuch as he possessed the secret and would have little difficulty in locating the mine, he would wait until the excitement had subsided before entering the mountains. After several months of continued search, the prospectors became discouraged and returned from the hunt. At last, when quiet prevailed, Father started out, preparing to remain but a few days. His search for the rock resembling a man was fruitless and soon his small store of food was gone. Weary, but not the least discouraged, he returned to Phoenix and immediately began to plan for other trips.

For a period of five years he searched the Superstition country for one object and one object only — the rock that looked like a man. If he could only find that one landmark, the rest would be easy, but when one observes the countless rocks and canyons and the hundreds of still unnamed mountains in that vast area of Nature's wrath, he can well imagine the difficulty of trying to locate a rock the size of a man. There is such a rock off the side of the famous Apache Trail. Many tourists today have observed it and have asked the prevaricating stage driver its meaning. If I were to disclose some of their answers to the curious and un-Western patrons of their particular transportation company, it might seriously affect future business along these lines — and that isn't being done, especially by an ex-stage driver, like myself, for I drove for the Apache Trail Stage Company in 1915-16-17.

Father was beginning to get discouraged. It was possible, he reasoned, that Jake Waltz had told him an untruth: he might have given him the wrong directions, preferring to keep his secret of the "Lost Dutchman" forever from the mind of the mortal man. He recalled the old German's story of having murdered his nephew, and in order to prove or disprove the story of the dying man, went to Agua Escondido. There he searched for and found the grave of the young German, just as it had been described to him by Waltz.

He had barely uncovered the earth when he came upon the skeleton of the German's nephew with the chain around what had once been the neck of a young man. The skull separated from the rest of the mass of bones and, hardened as he was to the wild and reckless life of early Arizona, he picked it up with a feeling of awe. He laid the skull aside and covered the balance of the bones with the soft earth. He returned to Phoenix with a feeling of renewed confidence in the deathbed story of old man Waltz. While passing through the town of Tempe, he stopped for a few moments to visit an old friend "Doc" Jones and left the skull with him.

Years later, an old prospector named Switzer came to me and told of finding the grave, and reopening it found the remains of the body with the chain minus the head. (This is just to disprove some of the stories that old Jake didn't have a nephew with him in the Superstitions.)

My father checked with a refining company in San Francisco and found that Waltz had shipped them, over a period of time, about two tons of very high-grade ore, for which he had received approximately two hundred thousand dollars. He also checked with Wells Fargo Company and found that he had sent a very large portion of this to a party in Germany.

In running down the highlights of Waltz's story, he also learned that the German had stopped one time at the Marlow Ranch (now 'Tex' Barkley's Quarter Circle 'U' Ranch) and secured some flour and bacon.

On another occasion, Waltz went to the Silver King

[Pinal] where he purchased a complete new outfit and a large supply of food. As he was making these purchases, he was accosted by an old acquaintance, Joe Fuggitt, who had worked with him in the Vulture Mine. Joe inquired if he had made a rich strike. Waltz told his friend that he was doing a little placering and, when Fuggitt asked if he could accompany him on the return trip, Waltz flatly refused to consider the proposition, explaining that there was not enough gold for the two of them. Joe left but returned shortly, for something he had forgotten, to find Waltz opening up a full bag of gold to pay for his supplies. (This was related to me by Joe Fuggitt himself.)

Doc. Jones of Tempe. He heard Dick Holmes' story and examined the skull of Waltz's nephew. Later his son, John "Johnny" Jones, settled at the ranch at Agua Escondido and became a firm believer in the Lost Dutchman Mine. (Source: Tempe Historical Museum)

Father's further investigation into the possibility of the location of the mine having been recorded, led [him] to Mexico City where he learned that there had been a mineral grant issued to one Pedro Peralta, consisting of approximately 21 miles square. The description of the grant coincided with the general description of the Superstition area.

His enthusiasm was considerably renewed at this disclosure, yet he had several more attempts to locate the mine without success. Not only could he not find the rock resembling a human figure, but he was likewise unable to locate the saddle in the mountains where Four Peaks were lined up as one peak on one side, and the large needle (presumably Weaver's Needle) on the other. It was true that he found a spot where Four Peaks appeared as described to him, but from this point he could not see even the top of the lofty needle [Weaver's Needle]. Always some mountain or mountains appeared to cut off the vision of one or the other of the landmarks.

He was not at all ignorant of the fact that his various trips into the mountains were being closely watched by those who had heard of his last conversation with the old Dutchman when he gave him the directions of the mine. Several times, when camped at some lonely spot, his keen eyes would observe not a few thin, spiral columns of smoke arising from the four points of the compass. He often thought of these unthinking followers and their campfires as compared with his fireless nights when trailing the Dutchman several years before.

* * *

[Brownie does not give much attention to Julia Thomas and none to the Petrasches; yet, they (Julia and the Petrasch family) are important figures in the Dutchman story. We know a serious dispute erupted between Julia Thomas and the Petrasch family on the one hand and the Holmes family on the other, and as a result there was little recognition by one of the other. It is perhaps for this reason that Brownie almost completely ignores Julia Thomas and the Petrasch family. The details of the three families and their feud are told in *Part I, The Golden Dream*, Chs. 13, 15 and 16.

As for others knowing of Waltz's gold, there were several besdies Joe Fuggit. At least 19 people can be documented as having known of Waltz's gold even at this late date. *Part I, The Golden Dream*, in Ch. 21 lists these 19 and references them as to source. Two of these sources are primary, one of these a document.

The present Quarter Circle 'U' Ranch was originally established by Matt Cavaness, who built the thick-walled stone house to withstand Apache attack. The ranch was later bought by George Marlowe, who built the first board house in that part of the country. The Marlowe family ran it as a dairy ranch in the 1880s. The ranch was then purchased from Marlowe's estate by Jim Bark and Frank Criswell in 1891, and named the Bark Ranch, a name that appeared on the original U.S.G.S. topographic maps of the area. In 1911 Bark sold his interest to his former partner Criswell, who then sold the ranch to 'Tex' Barkley. It is the last of the working cattle ranches of the western Superstition area, and while Marlowe's 'board house' burnt down many years ago, Caveness' original stone house still stands.

It appears that the mine of Jacob Waltz was hunted under various names until 1893 when the name 'Lost Dutchman' was reportedly given to the mine by James Bark (Ely, *The Lost Dutchman Mine*).]

Chapter 5
Dick Introduces His Son to the Search for the Mine

This chapter reveals one of the ironies surrounding the searches made by Dick and Brownie. What was supposed to have been the first of a series of trips by father and son to hunt Waltz's mine turned out to be their first and last trip together, and they never even entered the Superstitions on this outing.

I was barely sixteen years of age when, in 1908, Father asked me to accompany him on a further search of the great ore deposit of which I had heard so much. I was much enthused in having my father place so much confidence in me, and I can truthfully say that I learned much more on that first trip into the mountains, under his expert care and advice, than I have learned on any trip taken since. He was very particular that I become acquainted with the pitfalls of the mountains; [he] taught me how to pack a burro, how to get my bearings if lost, how to locate water in the summer and how to avoid being molested by the wild animals which stalk through the mountains' vastness.

We left Phoenix with our camp equipment packed on six burros. It required two days for us to reach the Verde River, where we camped over night near the spot used by Waltz when my father followed him years before. Father wondered, as he sat by the campfire that night, just how many prying eyes were focused on our every movement. He had told me before about his spying of old Jake that one winter night and he told me then that it was reasonable to believe that there were more than one pair of eyes concealed in the brush beyond our camp. Somehow the very thought that we were not alone made cold chills up and down my spine. I slept very little that night.

About midnight, I could hear the low rumbling sound of thunder in the mountains above us. Soon the sky became overcast; lightning flashed and the river appeared just a little higher as morning came on. When Father arose, he looked around apprehensively. He walked down to the river and stood for several minutes, watching the muddy waters. When he returned to camp, he casually informed me that the river was just a little too high for us to ford. So we remained there for another day and a night, during which time the river rose considerably and then receded as swiftly as it had come up.

While idling around camp the second afternoon, a man approached and asked Father about the wild life along the Verde. Father was suspicious from the start. It was evident that the man was not prepared to hunt or trap along the river as he carried nothing but a six shooter with him, together with a few camp utensils and some grub. However, the desired information was given to him, and he left after giving our camp a very thorough inspection.

The waters receded sufficiently for us to cross the river the following day and that evening we camped at Cottonwood Wash, about three miles north of Agua Escondido, where Father decided to remain for a full day. He explained that he was not in any particular hurry as he had two objects in view; one of which was to hunt for the mine and the other to teach me the art of being sure of myself on future trips into the Superstitions. I did not then realize that my father felt that he would never be able to locate the Lost Dutchman Mine.

The sun was just setting and Father was preparing the evening meal when we heard someone coming from the direction of a wash below our camp. Father casually picked up his rifle and motioned for me to follow him. We walked about a hundred yards down the wash when we ran face to face with our friend the trapper.

"You're a little off your trail, aren't you?", Father inquired of the surprised man.

"I don't think so," the man replied, somewhat sarcastically. He was well aware of the fact that we knew his purpose in being there.

"I thought you intended trapping along the Verde," said Father.

The man looked at Father and replied, "I believe a man is privileged to hunt wherever he damn pleases." With that he turned and walked away in the direction from which he came.

Father turned to me, "Son," he exclaimed, "you will be followed into the hills every time you make a trip. Sometimes there may be several people on your trail. Always remember to endeavor to shake them off, if possible. In the event that you can't get away from them, I would advise you to return to Phoenix. Remember, too, if and when you do locate the mine, don't let on that you have found it, for if you do, you'll never live to get out of mountains."

We returned to camp and finished preparing the meal. Occasionally, as we ate, father's keen eyes would scan the ever darkening wash. His attitude of watchfulness made me feel somewhat uneasy. He could sense my feelings and he endeavored to make me feel more comfortable by saying, "Son, don't be frightened of any person before you locate the property. They are harmless because they want to follow you to the mine. We are as safe here as we would be at home."

After that remark I felt more secure. I laid out my blankets, and placing a few sticks of wood on the fire, laid down to watch the stars twinkle in the heavens. In a few moments I was sound asleep.

The following morning we were up early. We ate our breakfast and had everything packed by sunup. As we left camp we back-tracked for some distance. I thought for a while that Father had given up the idea of a further search at that time and I was very disappointed. We had not traveled more than a mile when he turned sharply to the left. I knew then that he was headed for the Four Peaks country east of our previous night's camp.

That night we camped at Cane Springs and the following day my father took me to a spot about a mile and a half away and showed me where he had taken part in a battle with the Apache Indians when he was with the Government Scouts.

When we got back to camp, Father noticed strange footprints in the soft ground. Then he turned to me and grinned, "Our trapper friend paid us a visit while we were away," he said.

> [Believe me, the unwanted night time visitor is not something confined only to the past. Such men still prowl the mountains. Personal encounters with night-time visitors and other dangers of the Superstition Mountains are given in *Part I, The Golden Dream*, Ch. 19.]

We broke camp early the next morning and headed for Jeff Adam's cattle ranch on Long Gulch, where we camped for the night. After partaking of a excellent meal of venison and sourdough biscuits and beans, we retired to be up early again the next morning and on our way to "Hell's Hip Pocket."

"You are in what was once bad Indian country now, son," Father informed me. "This is where Al Seiber and his scouts killed a band of Apaches several years ago. Of course," he explained, feeling that I might not like the idea of being in Apache country, "those ornery devils are quite peaceable now and they don't do much harm."

I inquired about the details of the slaughter, becoming more interested since learning that we were perfectly safe from the merciless savages, who killed and ravaged the entire country under the direction of the noted Mangus Colorado, and the equally noted Cochise, famous and most brutal of Indian chiefs.

The first thing I want to do before I relate the story of the massacre, is to show you something of interest," he said. We walked down the side of the canyon and up on the other side. We came to a rather large cave and before going inside Father said, "Son you're going to see something that you have never seen before. I'm showing you this because I want you to know just what we had to go through in the old days. What you will see in there might startle you, but when you realize that we old-timers were constantly on the lookout to protect our lives from those murderous wretches, you'll understand that this little scene is not so terrible." With that we went inside the cave.

It required some little time for my eyes to become accustomed to the interior of the cave and when Father stopped suddenly some feet from the entrance and peered further in, I didn't realize that he was waiting for me to speak. I followed his gaze and slowly the sight unfolded before me. It was terrible. I gasped in amazement at the gruesome spectacle before me. There, inside the cave just six miles from Jeff Adam's peaceful cattle ranch, were human skeletons; not just one or two, but many of them scattered all along its floor.

I was horrified at the frightful scene spread before me. It presented a clear picture of a terrible battle that had taken place where now all was serene and quiet, except, of course, the eternal battle being waged by the elements and by the wild life which frequented these otherwise peaceful hills.

I could picture the terrific conflict of the Americans and the savage Apache Indians on that memorable day when Seiber and his band of brave scouts encountered the bloodthirsty Indians in the canyon below the cave; how the superior numbers of the white men drove the Apache braves into what they figured to be a natural stronghold.

I walked to the entrance of the cave, not daring to

Skull Cave some 40 years after the battle. The skulls and bones were lined up on the rock ledges for the photograph. Brownie said that when he was taken there by his father it still stank of human flesh. (Source: McClintock, *History of Arizona*)

look back. I experienced a peculiar sickening feeling in the region of my stomach and wanted fresh air — plenty of it. Father followed me out. When I looked at him, he was smiling. For a moment I felt somewhat ashamed for him to notice my weakness at my very first experience with death in actuality.

"My son," he said somewhat seriously, "I wanted you to see this sight so that you would understand what the pioneers of this country had to contend with constantly. The Apaches were blood-thirsty and treacherous savages. Women and children were murdered as were the men who protected them. They had no conscience, no sympathy for the whites. I was but a lad when they [were] doing their worst, yet so terrible were their crimes against all our people that I, too, gladly shouldered arms and fought them. I'll tell you the story of this massacre in detail later on, but now we must get back to the ranch before it gets dark." With that he walked ahead of me into the canyon and out into and over the hill on the other side to where our animals were tied.

When we returned to the ranch, we learned that our trapper friend had just passed through and was headed for Alder Creek, a couple of miles to the east, where he intended to make camp for the night. There was no question in our minds now but that he fully intended to follow us into the Superstitions and be right on the job in the event we were fortunate enough to find the Lost Dutchman Mine.

We remained at the ranch for four days, spending our time in robbing beehives and indulging in other harmless pastimes and at the same time keeping a rather close watch upon the activities of the trapper. It was evident that he did not intend to budge until we broke camp, so Father packed up our belongings and we traced our route back to Phoenix.

Although we accomplished little or nothing from a material standpoint, the trip proved to be a very liberal edu-

Billy Harwood, Brownie's stepson. As a boy listening to Brownie's tales of the frontier and the Lost Dutchman Mine, Billy became enamored with the Old West; in fact, this studio portrait was his favorite picture of himself.

Brownie and Billy were able to do what Brownie and Dick Holmes never could — search the Superstition Mountains together.. Billy and Brownie began a partnership which they hoped would last a lifetime. It is probably Billy who took the photograph of Brownie at Pipp's Cave (above Ch. 2). But, Billy's health turned cyclic and he could never hunt the Superstitions as they had hoped. (Courtesy: Holmes Family)

Indian Ruins, Superstition Mountains. Some of the indian ruins discovered by Brownie while searching for the Dutchman's house. (Courtesy: Holmes Family)

cation to me in the form of really learning how to rough it in a rough country. I had learned to care for the animals; to balance a pack upon the burros; to throw a hitch, which is very important in leading pack trains through the mountains; to select a suitable camp site and to avoid the deadly rattlesnake of which there were many, and of which we killed several during the many days on the trail.

This was the last trip of any length that my father ever made, for within a year, a leg injury, which he had received many years before while in the Government service as a packer, began to give him considerable trouble and compelled him to keep off the rough trails of the mountains.

Realizing his handicap, he told several friends to keep a sharp lookout for the rock house camp of the old Dutchman, and if found, to let him know its location, stating that this would be his most important clue to the discovery of the lost mine.

Many old Indian ruins and cliff dwellings were found as a result of this statement and various reports coming to Father caused quite a few trips into the mountains to verify them.

* * *

[The incident referred to in this chapter is the 1872 Battle of Salt River Cave (or Massacre as it is also called). Chief Nanni-chaddi and his people were cornered in their ranchiera, a two-room cave. After an initial attack, Nanni-chaddi was called upon to surrender. When he refused, the soldiers were ordered to fire as rapidly as they could, ricocheting their bullets off the ceiling of the cave which sloped at an angle of 45°. A separate detachment of soldiers climbed above the cave and rolled large rocks into the front of the cave; the rocks burst upon the trapped Apaches like shrapnel bombs.

When the soldiers eventually entered the cave, they found seventy-six men, women and children piled one upon the other, dead from ricochets and shrapnel, or crushed under rocks, plus about fifteen who lay dying. Eighteen Apache women and children survived, all wounded, to be taken prisoner (Bourke, *On the Border with Crook*).]

Chapter 6
Brownie Begins His Search

Brownie learns his father's health will no longer permit Dick Holmes to travel the mountains, and Brownie begins a solo quest. Ironically, a riding injury would eventually end Brownie's own searching, as the residue of a similar accident had ended his father's searches.

When I passed my seventeenth birthday, Father asked that I come into his bedroom one morning as he had a very important message to convey to me. I walked into the room feeling strangely sorry for him as he lay upon his bed.

"Son," he exclaimed in his rather good-natured tone, "I have come to the end of my journey so far as the Lost Dutchman is concerned." He smiled as he saw the tears forming in my eyes. "Now don't feel that way about it son, I'm a pretty good man yet, but there's times when this old leg of mine gives me a lot of trouble and I must stay out of the mountains."

"Won't you ever be able to go with me again, Father?" I asked, knowing full well that the question was superfluous, but I had to say something and that was the only question I could think of at the moment.

Dick Holmes at about the age he had to abandon his physical searching for the mine. (Courtesy: The Holmes Family)

"No, son, I can't travel the trails with a stiff knee, so I'm going to let you do it for me. We'll form a sort of corporation between us. I'll be the director and you will be the field man and remember, what I say goes. You are old enough now to be on your own and I want you to continue the search just as I have conducted it in the past."

He then gave me the instructions, some of which appear in this narrative from time to time.

"When do you want me to start?" I asked with a feeling of manly importance.

"Whenever it's convenient with you," he replied. "And remember, son, don't be too hasty and, if you do locate the rock house, do not look for the mine, don't even attempt pacing it off as I have described to you, but return to Phoenix immediately so we can secure the assistance of authorities to protect us — then we'll go back and locate the property and record it before beginning operations. Always remember, too, that there are eyes watching your every movement while in the hills, you are never alone. One word of exclamation from you or any outward indication from you that you have made an important discovery will result in trouble. With that, he dismissed me.

Brownie (left) and his father: Although Dick Holmes could no longer travel the rough terrain of the Superstitions, he and Brownie continued to camp and pack together in less inhospitable terrain. Near Wickenburg circa 1917. (Courtesy: The Holmes Family)

You can well imagine how I, or any seventeen year old boy, would react to such an important assignment by his own father, whom he had worshiped from babyhood; especially when the assignment concerned the search for a fantastic mine of wondrous wealth hidden somewhere in the vast interior of a range of mysterious mountains, the outlines of which could be seen from Phoenix.

[This is apparently how Dick and Brownie Holmes worked for many years. Dick Holmes believed that he had been looking in the right part of the mountains, and that all he had to do was find the last few clues which would precisely locate the mine. So Brownie was sent in to look for those clues. But the last clues were all fugitive, and they disappeared before they were ever located.]

It is needless to say that I lost no time in making preparations. I succeeded in getting a pretty fair outfit together, which consisted of bedding, tents, cooking utensils and food. This large quantity of necessities required seven burros to carry it. I asked for and received permission to let a boy of my own age accompany me on the journey. With a skeleton of the route we were to travel, accompanied by a feeling of youthful exultation, we found ourselves on our way one bright late summer morning to relocate the Lost Dutchman of the Superstitions.

After seven days of travel across the Verde and the Salt Rivers, up through mountain passes and canyons and over hills, we finally found ourselves camped at Indian Springs.

We camped at Indian Springs (🚶) for about three weeks, familiarizing ourselves with the many canyons and other landmarks and, on several occasions, exploring the country well inside that awe-inspiring and mysterious range of mountains.

My friend seemed to enjoy running down the several species of wild life, whose habitat was in that vicinity, rather than look for hidden treasure. Several times I would call his attention to some peculiar formation of rock high up on a mountain and his answer would be that a coyote or deer or some other animal was stalking through the underbrush. One morning he came into camp with a Gila Monster dangling on

---

🚶 : Trip 42, Peter's Trail to Charlebois Spring.

Indian Springs.

the end of a piece of string, informing me that he was going to take the deadly creature back to Phoenix.

One night as we finished our meal of beans, biscuits and jerky, I began to experience a strange feeling of hidden eyes watching.

"I looked over at my companion to see if he felt the same way and I noticed him gazing across the arroyo, the other side of which were thick bushes of oak and mesquite. My gaze followed his—yet I could see nothing except the ever-darkening desert flora. Once my companion looked at me in silence. There seemed to be an electric connection between us in which we sensed the same uncomfortable feeling of unpleasant visitors.

I remembered what my father had told me on a previous trip, when he said that we were safe from harm until we found the hidden mine. That thought gave me a renewed courage and, in the clearest and bravest tone of voice I could possibly muster up under the circumstances, I startled my companion half out of his wits by bellowing, "Come on over pardner and have something to eat!"

After shouting to who I felt surely was some person or persons of our acquaintance in Phoenix, if our intuition was correct in surmising that we were being spied upon, I watched the thicket closely and thought I saw a dark object move off away from our camp.

The following day we started back to Phoenix.

* * *

[Eventually, Brownie and Dick Holmes abandoned their immediate search area which was in the canyons around Weaver's Needle. They expanded and extended their search area until they eventually hunted throughout the entire Superstition Mountains from Tortilla Flat in the northwestern part of the mountains to Randolph Canyon in the southeastern part of the mountains. When they failed to find the Rock-That-Looks-Like-A-Man or the Rock House, they began to reconsider the idea that the mine was in the Superstition Mountains.

Waltz did not give a name to the mountains where the mine was located when he told his tale to Dick Holmes; and Rhinehart Petrasch, who knew the name of the mountains, was not about to tell Dick Holmes (See: *Part I, The Golden Dream*, Ch. 12). Waltz did not even tell Holmes whether the mine was north or south of the Salt River.

Eventually, Brownie and his father widened their search to include the Four Peaks area to the north of the Salt River. This area is more heavily mineralized than the Superstition Mountains and seemed promising. Brownie did find an old abandoned Mexican mine.]

Chapter 7
The Tree Stumps and a Trip with a Single-Blanket, Jack-Ass Prospector

The title says it all. This singular expedition was one of Brownie's most productive, colorful and memorable.

During my next trip, I camped at Marsh Valley in La Barge Canyon after nearly a month of searching without finding anything to encourage me.

I was anxious to get out of those desolate hills and return to Phoenix. I would not have taken this particular trip had it not been for a new clue, which I was desirous of running down and which had a very important bearing on the possible location of the rock house. A prospector had mentioned having seen a rock house, partly caved in, near the upper end of the canyon. I did not place any real faith in his story, because I was of the opinion that the lost mine was not in that immediate vicinity. However, one must not overlook anything that might throw some light on the location of the concealed ore deposit. I hunted diligently for the rock house for several days without success, finally confining my search to La Barge Canyon.

I planned to return to Phoenix after one more day of searching. As a matter of fact, I decided to put in my last day in an endeavor to find some trace of the vein, which the old Dutchman said had been partly exposed in the canyon and which he carefully covered over with rocks and sand to lessen the possibility of detection.

La Barge Canyon had experienced a flash flood some days previous and I felt that there might be some indication of exposed ore as a result of the devastating onrush of water.

I left the canyon and climbed over a thickly wooded hill, and was following along the top of the hill leading toward the head of the canyon, when I made a discovery that made my heart stand still for a moment. Here and there were tree stumps! Some one had sawed down the trees for a particular purpose and the purpose, I felt sure, was to cover the shaft of the Lost Dutchman of the Superstitions!

I cannot begin to describe my feelings after making this most important discovery. I had uncovered other clues in this vicinity and I was sure the mine could not be far away. I started back to Phoenix to impart this knowledge to Father. I even rehearsed the manner in which I would deliver the welcome news to him. I would tell him, casual like, that I had found the place where old man Waltz had sawed down the trees to cover the mine. Then I'd sit there and watch the reaction of my nonchalant announcement to Father.

Trees Stumps — one of many old, stooled out mesquite stumps in the Horse Basin Country. Are these the possible clues to which Brownie was referring? See: *Part I, The Golden Dream, Ch. 22.*

As I made my way to Phoenix, the whole world had a very different look to me. I could see my father and mother dressed in regal splendor—with more gold and money than any other person in the world, as a result of discoveries made by me, their own son. I could see the army of workers, guarded by uniformed and heavily armed men, as they took out the millions and millions of dollars in gold from the shaft of the Lost Dutchman Mine.

When I returned to Phoenix and told my story of the locating of the stumps, Father did not seem to think the discovery was of any great importance. As a matter of fact, he felt he could have made the same discovery without much effort. It was true, he reasoned, the location of the stumps indicated the close proximity of the mine, yet the old Dutchman was so intent on concealing the thing that he might have dragged the logs a considerable distance. The canyon which the logs might have been carried into had been combed from one end to the other by both of us. I examined the hill sides thoroughly and the rock house was not to be found.

I was somewhat disappointed in finding that Father had treated so lightly what to me had been such an important

Following the Dutchman's trail. A family outing panning for a trace of the Lost Dutchman or Lost Spanish Mines of the Superstitions.

discovery. He pulled out the map he had made from Jake Waltz's description and placed a small circle at the place described by me as the tree-stump hill.

> [Whether or not Dick Holmes ever drew a map is problematical. Some who knew Brownie say there never was a map, while others who knew him claim there was a map.]

I looked over the crude drawing and observed that most of the notations on it were within a certain area. After studying the map for a moment, I could see why Father was not enthused over my last discovery. Everything seemed to center around two canyons, either one of which might conceal the Lost Dutchman Mine; they were both within a limited area, however, and somewhere within that area the rock house must lie, completely hidden from the eyes of man.

An old man, a pioneer of Mesa, informed my father that he knew of a rock house in the Superstitions that answered the descriptions of the one the old Dutchman once occupied. He agreed to accompany me in search of the house and so one morning we started out with a pack train and provisions for a month or so or more. The fellow was about seventy years old and was very active for his age, having done considerable hard work all of his life. This fact made me feel a little more cheerful regarding the trip as I was not very enthusiastic about having an aged man on my hands when Father first informed me that the man was going with me on the trip.

It did not require much time for me to learn some of my companion's peculiarities. We had not gone very far when he began to quote Scripture. In fact, most of his conversation concerned religion and he dwelt rather heavily upon the evils of acquiring wealth. I was very much impressed with the old fellow's piousness and was constantly on my guard lest I emit some sort of cussword that might offend him.

I had been brought up in a good family, yet I was almost constantly coming in contact with a rather rough element and the apparent goodness of this old man made me feel somewhat ill at ease. I openly admired his faith in an everlasting hereafter and secretly held him in awe, because of the fact that after seventy years of a very active life in the raw and generally lawless west, he was rounding out his remaining days with such a clean heart and pure soul. I felt sure that when he stepped out of this world into the celestial one above, he would be given the highest reward they had to offer.

"My boy," he said after we had traveled a while in silence, "I feel as if I were doing something sacrilegious in taking you up in those mountains to search for gold. Since the beginning of recorded history, gold has been a curse upon mankind. It has truly been the instrument of the devil. All of the disastrous wars of history had their beginning because of the lust of certain people for the unholy metal. Men will murder and steal because of it, and women will go to the bottomless pits in hell to search for it."

I wondered at the time why he was going to so much trouble to assist in the location of the lost mine, but feeling that it might be another of his many efforts to aid mankind, I refrained from asking him why he should be so anxious to help find a deposit of the devilish, destructive yellow metal.

We arrived at Boulder Canyon and made the laborious journey half its length when the old gentleman became confused and seemed to lose his bearings. I realized, that after a lapse of several years, it would be an easy matter for a person to forget the location of the many and varied landmarks of that vast range of nameless mountains, so I refrained from hurrying him along, although at times, I became impatient to a marked degree.

We stopped to rest at a spring I had discovered on a previous trip. The old fellow did not make an effort to assist me in preparing the meals, but he was the first to sit down after they were prepared. With a solemnity befitting a wearer of the cloth, he offered profound and sincere thanks for the privilege of participating in such a bountiful repast such as was placed before him.

After partaking of the "bountiful" meal of hard biscuits and cold ham, washed down with an occasional swig of spring water, we resumed our journey.

"Do you remember this part of the country?", I asked after we left the canyon and reached the brow of a steep hill.

"I don't just remember, son," he returned, "but have faith, and all will be well." With that he read me a passage out of a well-worn Bible he had resurrected from some portion of his well-clothed anatomy.

We [were] descending from the top of a sharply sloped hill when the burro, which the old man was riding, stumbled on a rock and as a result, the saddle slipped, throwing the old fellow rear end first into the only bed of cactus on that mountainside. As I remember, he had his Bible in his hand at the time, reading half aloud something about Moses saying, "Honor thy father and thy mother."

The thing struck me so funny that I laughed. I realized the seriousness of his predicament, however, and immediately dismounted to assist him in extricating himself from his painful resting place. You can well imagine my shock and surprise, when he let loose with the most artistic stream of blasphemy I had ever heard. He completely ignored my offer of assistance. He had eyes only for the poor, defenseless burro.

"You damned—————, " he exclaimed with fervor, "I'll cut your throat as soon as I can get out of this damned mess."

He made no effort to get up—just sat there, calling every burro in the outfit every name he could possibly think of, back to their immediate ancestors and thence back to the very beginning of the species.

My temporary shock turned to undisguised admiration as he took up different subjects with which to give vent to his extreme anger—the saddle, the rocks, the cactus, the Superstition Mountains, the Lost Dutchman Mine and the idea behind the journey; each received its proportionate share

Prickly Pear Cactus — the source of the old prospector's discomfort.

of his uncontrollable wrath. He analyzed each and every one in its proper turn, dexterously maneuvering his unusual and most remarkable vocabulary of cusswords in such a splendid manner that there was scarcely a repetition of a single word. It was really the greatest masterpiece of extemporaneous swearing I had ever heard.

When I finally induced him to descend to the surface of the earth and get down on all fours so I could begin the delicate operation of extracting the cactus from the rearmost portion of his well pierced anatomy, he started from the very beginning and repeated his voluminous and very descriptive assortment of profane and obscene words, which are not found in any dictionary.

To those who have never had the experience, permit

me to say that a cactus thorn, regardless of what species, and there are many in Arizona, hurts far more coming out than it does going in, because of the minute hook on the end of each needle-like point.

When the old man would pause for breath, the neighboring cliffs would re-echo some choice expression, thereby keeping up the continuity of the seventy years of stored up cuss words. I held the old man in a different kind of awe, from that time on, and as a result have not taken up any kind of swearing, because I fully realize that I could never reach the height of perfection so ably demonstrated by that pious gentleman of three score years and ten.

I have often thought of the old fellow and of his going from one extreme to the other and the only way I could possibly figure out why he gave up his religion for an hour was that there were no words in his little Bible that could truly express the feelings of one who had the seat of his trousers, and beyond, full of cactus thorns.

We spent another ten days in a fruitless search for the rock house, during which time we paid unusual attention to the burros, examining the critters at various intervals to make sure the cinch was tight.

Strange as it may seem, my heretofore pious friend kept his Bible in seclusion. He also refrained from delivering lectures upon unholy things in general. Several times, I observed him looking at his burro somewhat contemptibly and I figured that he was still contemplating on cutting the varmint's throat.

At last the old fellow confessed that the lapse of many years since he first discovered the rock house had impaired his memory quite extensively and he was unable to get his bearings sufficiently to enable him to direct me to its location, so we packed our outfit and returned to Mesa.

We stopped in front of his house and relieved the animals of his portion of the luggage. I bade him good-bye and started on my journey to Phoenix. Just as I got under way, the old man walked up to the burro, who had long since forgotten the catastrophe of several days before, and gave him such a terrific kick for his seventy years, that it would have done credit to any husky football player of today.

He passed away about a year after our trip into the Superstitions and I can well imagine him resting in sublime peace where there are no burros, slipping saddles or lost mines and especially no cactus-studded cushions.

* * *

[I have had the privilege of knowing one of the last of a dying breed — a true single blanket, jack-ass prospector, Verne Harkness. Back in the old days, the '20s and '30s, he prospected the southwestern deserts with two 'Shetland mules'. I've seen him look at a hillside, spot the location of a vein, tell me its depth and whether it was worth mining or not — where I, a greenhorn, could see nothing, and I mean nothing.

One day up at the old OK Mine in the Dale Mining District in the Mojave Desert, Verne took me out to show me some of the old mine. He pointed at several places on the surrounding hills and ridges, asking if I could see a vein. I could not. After a bit, I think Verne thought that I thought he was just pulling my leg. So we walked down a hill, ostensibly looking for artifacts, when Verne turned and pointed me at a small ridge he had earlier pointed out from above as having a vein, a ridge that to me did not look any different from the rest of the Mojave. But, from down below you could see where Verne had used a back hoe to dig into the ridge and there in the exposed dirt and rock even I could see a good sized vein.

Hugh d'Autremount, a mining engineer, laughed when I told him about Verne, and said one of the first things he had to learn as a young buck was to take the old time prospectors seriously. He had learned the hard way. For it seemed when he was younger he put a great deal of stock in his tests and books, having little use for the 'old-time unscientific prospectors.' But, in those days they were still around and a lot of his clients would employ them, plus professional geologists and mining engineers. He said many is the time he would scoff at a prospector's conclusions, just to be proved wrong. He realized his clients were not paying him to be wrong, and began to listen with respect.

There is a big difference between a man prospecting and a prospector, a big difference. I've known lots of men who prospect. I've known one, maybe two, prospectors. Waltz reportedly said: "That no prospector will ever find my mine." — which would have to mean, at least to Waltz's eye, there are no surface indications of the vein.]

Verne Harkness up at his diggins in the Mojave back in '87. Verne was trying to re-open the old OK Mine in Dale Mining District a few miles east of the town of Twenty Nine Palms. It was in the Dale district that Hugh d'Autramount first mined. I can attest that Hugh wasn't exaggerating when he described the country as so dry even the jackrabbits carry canteens.

Chapter 8
The Mountains and Their Place in the Pima Legend of Creation

This chapter is a compilation of stories about the mountains and their place in legends of the Pimas, a Native American tribe. For some reason, the early Dutch Hunters felt compelled to give their version of the Pima creation and flood legends. Jim Bark did the same in the 'The Bark Notes', as did Sims Ely in his book.

The decided contrast between the summer and winter in these awe-inspiring hills is worthy of note. The summer months portray them as being on the very brink of Hell. No two formations alike, those crazy indescribable, unnamed peaks and canyons present a picture of utter desolation.

In the winter and early spring the many, many rains change the panoramic view, and make a veritable garden out of certain portions of the vast range. In six weeks, grass and poppies growing abundantly in the valleys, produce a picture no artist can imitate, while the hills which surround them

A rain camp. Everything — even our meager fire, covered by tarps.

remain barren, rocky slopes for the most part, with only here and there an occasional green patch of mesquite and ocotillo and other wild growth.

Many peculiar and sometimes unexplainable incidents can take place in the Superstitions, that might well have their grotesque and incomprehensible effect upon an uneducated and inexperienced person or superstitious Indian. The nights, unless there happens to be an electrical storm raging or a high wind blowing, are so still that the least noise becomes greatly magnified.

I have on many occasions heard the sound of trains traveling twenty-five miles to the south. I have laid on my bed at night and actually heard music, which sounded very much like the Salvation Army Band playing on the streets of Phoenix, forty five miles away. One person, whose integrity I have no cause to doubt, informed me that he would have sworn that he heard someone step on the starter of an automobile and drive away; this from the remote Charlebois Canyon.

Tex Barkley, whose stories are always just a little more astounding than most, has said that he could hear the Mexican National Band after a siesta in the government park at Mexico City. Of course, 'Tex' would not exaggerate anything of that sort and in order to retain his friendship, which I cherish to a marked degree, I am compelled to believe his story!

At times there will come a very high wind that will touch only the highest crags and pinnacles. When this happens, a very weird sound is produced, sometimes like the soft strains of far-away music, and at other times not unlike the piercing cry of a person in mortal pain, followed by inarticulate moaning, and then by that unpleasant, awesome silence. The sounds are more predominant in Boulder Basin than elsewhere, in my opinion.

There is constant erosion in such a rough country and when a large boulder is loosened and rolls down the mountain-side or over a high cliff on a still night, the echo of the accompanying roar is increased a hundredfold. This is very disconcerting, especially if one happens to be alone. The nights, with the magical, fantastic lights of the heavens and the moon waning beneath the horizon, make it seem all the more uncanny, when without previous warning, the blood-curdling cry of the mountain lion breaks the stillness, like the cry of a woman. When this occurs, one will almost invariably throw another stick of wood on the campfire to thaw out the cold chills which have a peculiar tendency to creep up and down the spine.

It takes only one animal to start the bedlam of exasperating noises. The owls, which abound in the many caves and hidden recesses of the mountains, take their turn along with the coyotes, to be followed by the melancholy wail of the mourning dove. With the alternate deathly silence and the distorted, wild and incoherent display of animal confusion, a person must not let his imagination get the better of him. If he does, he is most certain to spend a very uncomfortable night.

The many pranks which nature plays on the unsuspecting person are enough to make an indelible impression upon his memory. On some still nights, I have seen the blaze of my campfire reach for many feet in a thin column and then suddenly flatten out along the ground, covering a considerable area; the flames licking objects in every direction like some fiery demon one reads about in fairy tales.

The high winds blowing from the lowlands to the west are responsible for this peculiarity. One moment the blaze is sucked high into the air, because of a semi-vacuum created when the wind strikes the neighboring cliffs, and then the down draft, which always follows, causes it to creep out over the ground. All of these phenomena are laid to the evil spirits, by the Indians and others of a superstitious nature.

The Pima Indians hold the Superstition Mountains in absolute awe, for, according to their legends, these hills were the birthplace of man. It was here that Montezuma was created, where the great flood took place and where the Pima race originated. A brief synopsis of their legend is given here just as it was once told to me by an old man of the tribe.

THE PIMA LEGEND

One time many many years ago there lived atop the Superstition Mountains a beautiful maiden, who was good and kind to the River People (Pimas) and who aided them in time of famine. She always had plenty of grain and many seeds and when the waters of the river were low, she would keep the River People from becoming hungry and so they prospered and were happy.

One time, when the beautiful maiden lay asleep, a rain came up and a drop of water fell upon her navel and she became pregnant and bore a son. This son was named Montezuma and he became the ruler of the Pima tribe.

Montezuma built a great house (Casa Grande), the lumber of which was taken from the mountains far away. The Rain God and the Wind God were his servants and they did as they were bid. The Sun also was the servant of Montezuma and he made the crops grow after the rain fell upon them.

One day Montezuma became angry at the Rain God and the Wind God and he banished them from the land. After they were gone, the crops would not grow and the Sun became very angry and scorched everything that was growing and a famine came upon the people. He sent his servants to find the two Gods and bring them back and when the Gods returned, there came a great rain, which lasted for many days and which drowned many people.

Montezuma went atop the Superstitions and waited for the waters to recede. The water came far up the mountain side and the foam marks could be seen along the sides of the mountains, and they were afterwards called Foam Mountains, by the Indians. After the waters went away, Montezuma made many men out of clay and sent them in different directions and later when he wanted to find out how they were behaving, he sent messengers to find them. They returned to report that some of the tribes could talk a strange tongue and were hostile (supposedly Apaches). Montezuma became angry at the people with a strange tongue and declared them evil, and were not to be considered friends, and so from that time on the Apache has been a bitter enemy of the Pima Indians.

The remains of an old (former?) rock house located in Needle Canyon that has intrigued more than a few Dutch Hunters. (Courtesy: Superstition Mountain Historical Society)

A portion of the Pima legend described the fury of the Fire God and how he punished those who were evil and would not obey. The old Indian's description of some of the methods of punishment meted out to the unruly coincides with some of my own experiences during the severe electrical storms.

On two occasions I have seen the atmosphere so charged with electricity that there was a continual crackling and hissing noise of several minutes duration, accompanied by a peculiar glow and unusual odor, which reminded me of accounts I had read of the Aurora Borealis. Once when this occurred, I attempted to mount my horse. When I touched his neck, I received such shock that the sparks actually leaped from the body of the horse to that of mine. This experience made me feel ill at ease, but later the cause was explained to me as static electricity.

* * *

[Jim Bark has also commented about the voice of the Superstitions, the loud resounding boom to which Brownie refers. He could never identify its source, even though at times he and his cowboys tried repeatedly to find the source, even camping out in the mountains so they could get on the trail of the sounds immediately. They were always unsuccessful in finding the cause of these sounds.

I have sat in the old Goldfield District over a mile from the western end of the mountains, and listened to these booms. While the sounds appeared to experience a near hiatus for many years, they have most definitely returned. On our last trip the mountains they were very active, and we heard these booms several times a day, day after day — and in the mountains they can be quite impressive, and to some can be genuinely disconcerting. I have heard of people who, having heard these sounds, have almost fled the mountains in terror to escape whatever entity was making them. Their source remains a mystery. However, I have wondered if it was not these loud booms that may have contributed to the Apache indians' belief that these mountains were the home of the Thundergods. Consider hearing loud sounds not unlike thunder emanating from some unknown place in the mountains on a bright day without a cloud in the sky.

Brownie's version of the Pima Flood and Creation legends is much abbreviated and differs in some ways from the more accepted accounts. For example, in most recountings there is no mention of Montezuma. The principal protagonists are: Juhwerta Mahkai, Ee-ee-Toy and Toehahvs. However, Brownie's tale agrees very well in broad terms with most other versions, and all the accounts agree that Slanting Mountain (Superstition Mountain) figured prominently in the Flood legend. (Lloyd, *Aw-aw-tam Indian Nights: Being Myths and Legends of the Pimas of Arizona; Shaw, Pima Indian Legends*).

The names Montezuma and/or Aztec are ascribed to most sizeable works or ruins in the Southwest. As for Casa Grande, which means 'Large House' in Spanish, Casa Grande National Monument is located about 40 miles southeast of Phoenix. It is the largest single surviving dwelling ruin in the American Southwest.]

Casa Grandes 1864

Chapter 9
John Chuning and the Lost Deering Mine

What follows in this chapter is one of the most intriguing and best verified of all the stories or reports of others finding the mine of Jacob Waltz.

The time came when my father suggested that I go to Tortilla Station and there endeavor to make friends with an old man named John Chuning, who, it was said, had some information regarding the lost mine. I was told to invite him to accompany me on the search for the Lost Dutchman. Father informed me that Chuning was at one time a shift boss in the old Silver King Mine, not far from the Superior; and a friend of his, Joe Deering whom I have previously mentioned, had come to Silver King with some very rich specimens of gold ore. He asked Chuning to put up the money for a first-class outfit and go with him to the mine.

Chuning agreed to accompany Deering to the mine and also to invest several hundred dollars in a complete outfit provided Deering would wait about a month. This Deering agreed to do and, in order to pass the time away more quickly and also to assist as much as possible in the finances of the venture, he went to work in the Silver King mine.

On his third shift underground, a rock slide occurred and both his legs were terribly mangled. One of them was amputated in an effort to save his life and, because of a lack of proper medical facilities, blood poisoning set in. He died during the second operation performed a few days later.

Before Deering passed away, he called his friend Chuning to his bedside and endeavored to give him the directions to the mine. Chuning learned enough to give him a fairly accurate description of the mine's location; however, he made no effort to look for it until after Waltz's death. When he began to hear rumors of the rich ore deposit in the Superstitions, he surmised that it was the same property which his friend Deering described to him.

> [Although Deering was a somewhat ephemeral historical personage, in that it has been difficult to locate his name on mining claims or in census reports, Deering was an historical person who did die in a mining accident as Brownie describes. The details of the Deering story are laid out in *Part I, The Golden Dream*, Ch, 9.]

Prospectors in those days were no different from those of today. A man may come into a rich strike and, unless he has several hundreds or thousands of dollars worth of ore on his person, no one pays particular attention to him. Chuning felt that his friend's strike was about the average because the ore which Deering showed him, although it was exceptionally high-grade, was of such a small amount that it would not pan more than three- or four-hundred dollars in its entirety.

The Silver King where Deering worked and died. (Source: Report of the Silver King Mining Company, 1882)

When he learned of the passing away of Jacob Waltz and of the many stories connected with the old German, Chuning realized that he had passed up one of the major opportunities of his career as a mining man. He procured an outfit and went to Tortilla Spring where he set up a permanent camp.

I found Chuning to be a pleasant fellow of about 65 years old. He was a wiry man in spite of his age; tall of stature, slightly bent with piercing grey eyes that seemed to look right into one as he talked to them. Yet there was a slight twinkle in them which indicated a kindly disposition.

As I rode into his camp, he was just clearing away the remnants of his evening meal. He looked up a little startled, I thought.

"Hello, there, " I said, "You're John Chuning, aren't you?"

He walked over to me as I dismounted from my burro, "Yes, " he said, "and who might you be?"

"I'm George Holmes from Phoenix," I replied.

"Not Dick Holmes' boy?" he inquired.

"Yes, sir."

"Well come right over and I'll fix you a bite to eat," he exclaimed, good-naturedly.

I protested, declaring that I could fix my own meal, inasmuch as he had finished his and cleared everything away.

"Listen, son," he remarked, "no one can say that they ever came into John Chuning's camp at mealtime and didn't get a bite to eat. I'll have something fixed for you in a jiffy."

With that he set about making over the coffee and frying some bacon. As he worked, he asked about Father and was surprised to hear that he was unable to hit the trail again. "I only met him a couple of times, son, but what I saw of him those times and from what I've heard of him at different times before and after meeting him, I think he's a real man."

John Chuning must have known my errand, but if he did, he never let on. He treated me as if I were just a casual visitor and would probably move on the morrow.

The following morning, I insisted that he prepare breakfast from my own store of food, but he would not have it, declaring that I was his guest. It was only after I told him that I might remain there for a couple days that he finally consented to share the food equally. Even then, as is characteristic of all old-time Westerners and more particularly prospectors and mining men, he did not ask me my mission, feeling that if I cared to, I would tell him of my own free will.

After the breakfast was over and the things were cleared away, I broached the subject of my visit. "Mr. Chuning," I said a little hesitantly, "Father asked me to come and see you about having you join me in looking for the Lost Dutchman. We have some very accurate information

"When Deering got to the creek he cut a cross on a black willow tree" which I have seen many times. *(Illustration: Randy Squires)*

that, coupled with what you know, might aid both of us a great deal."

His immediate reply convinced me that he had known or strongly suspected the reason why I was there. "No son, I feel that I'm on the right track myself, and [you] know how we old prospectors are — we don't like to let anyone in on a good thing if we can locate it alone." He readily observed the look of disappointment in my countenance. "Now don't misunderstand me," he went on, "I don't know of anyone that I'd rather have in cahoots with me than Dick Holmes or his boy, but you see I've been up here a long time and I've located a lot of valuable clues. I'm going to find that shaft one of these days, and when I do, I intend to own the thing lock, stock, and barrel."

I spent several days with him, during which time I learned that he was far from being on the right track. John Chuning was looking for an open shaft and the Lost Dutchman mine was covered over with 6 ft. of timber.

Chuning confined his search to the high cliffs in the vicinity of Tortilla Springs. He built ladders to climb up and down several sheer cliffs, and some of those ladders are still there. Before Mr. Chuning passed away in 1922, he told me some other things that Deering had related to him. Deering had, according to Chuning, built five markers from the mine to the big canyon to the south and had made them all alike. They were made of a long rock, standing upright, supported by four smaller rocks at the base. When he had got to the creek he cut a cross on a black willow tree which I have seen many times. That tree has since been washed away, but I have found three of the five markers Deering erected. The photograph of one of these markers is shown.

[No photograph accompanied the manuscript. What happened to it is unknown. In *Part I, The Golden Dream*, Ch. 9 is a photograph of a stone marker or monument which may have been one of Deering's. It is one of three such "markers" known, all on different mountains.]

After this fruitless trip to enlist the aid of John Chuning, I decided to make a more systematic search. I therefore began my renewed quest by starting out at Superstition Mountain, the main mountain of the range, and gradually working north and east, thoroughly examining each canyon as I went.

My various trips throughout the country always ended near the point where I had discovered the tree stumps. It was unreasonable to think that the old German would cut logs and drag them into some other canyon. To have done that would have required many months of hard work, yet I reflected after my first enthusiasm had died down, I must find the rock house in order to locate the shaft.

[It is obvious that by this point neither Dick nor Brownie Holmes had any specific idea as to the location of the mine. They had entered that period of their search where they roamed far and wide, including going north of the Salt River to explore in the Four Peaks area. We now know how their search may have been, at least partially, thwarted.]

I confined a portion of my time in the hills to search for the remnants of the Peralta party, but without any degree of success, except that I did find some old sandals and a pair of old bits in a cave on one of the canyon walls. Whether these might have belonged to the party or not, I have never been able to determine. But their findings spurred me on to more prolonged searches.

John Chuning informed me that a man by the name of Silverlock and a partner passed through the mountains several years previous about the year 1900 and, not knowing anything about the running fight between the Indians and the Mexicans, found numerous small pieces of rich gold ore. They continued their search and ran across several piles of the same rich ore.

It has been said that these two men took their find to Phoenix and received approximately forty thousand dollars for it. I have located the spot where the fight ended but, so far as I know, no more gold has been found there.

Up to this time, I had searched most in the area south of La Barge Canyon. Incidentally, while on one of these trips, I got my first glimpse of the dwarf deer. He resembled a jack rabbit with antlers no bigger than the palm of a man's hand. He was not over eighteen inches high and would not have weighed over twenty pounds. His body was mouse colored with dark spots about the size of a dime covering it.

I have seen many dwarf deer since that first tiny buck. On one trip, I observed almost half a hundred in one bunch.

When they saw, or scented me, they would scamper off in every direction through the rocks.

Later, after looking them up in the library, I learned that they are natives of South America and related to the Dik-Dik. Others have seen these animals in the Superstitions, but many people seem to think I am telling a customary western yarn when I say that I have seen a great number of them.

I have seldom made a trip into the Superstitions that was not indelibly impressed in my memory through some experience. One time I found parts of a human skeleton that were so old I did not go to the trouble of reporting the find to the authorities. The skeleton was lying on the side of Crab Canyon, so named because a member of the party which I was guiding when we discovered the canyon was so irritable that we did him the honor of naming the gorge after him. It was dedicated, placarded and recorded as Crab Canyon amid elaborate ceremonies in the presence of this crabby individual. That name will remain unchanged forevermore.

Once when camping in La Barge Canyon, near Marsh Valley, I began preparing the evening meal. The sun was just setting and as I had traveled quite a distance that day. I was very tired and hungry. After the coffee had boiled and the beans had been sufficiently warmed, I sliced out enough bacon to round a good meal. I had no more than placed the bacon in the skillet when I heard a noise coming from the top of the adjoining hill and looking up saw a man with four burros coming down the hill trail.

In customary western style, being pleased with the thought of having a companion for the night, I added more bacon to the skillet and set out another plate and cup, at the same time gauging coffee to make sure there was enough for two. Satisfying myself that everything would be in readiness for the reception of my welcome visitor, I waited his descent from the side of the rocky hill.

He seemed to be coming straight for my camp and as he approached to within hailing distance I shouted out an invitation for him to "Come and get it."

He made no effort to halt but drove his burros right past my camp without one look in my direction. That was the first and last time I ever saw him. I ate my supper alone.

There are peculiar people who go into the Superstitions; most of them being eccentric prospectors, who having spent many years alone in the mountain vastness, do not particularly care for the companionship of man. Some of them will walk miles out of their way to avoid passing another prospector's camp. There are others that possess the insane feeling that they are just about to locate the Lost Dutchman mine and when they see others in the mountains, they feel that they are there for the sole purpose of robbing them. Men of this type are actually dangerous and at times they are very liable to take a pot shot at some unsuspecting person.

* * *

[Brownie's reference to two men working an area on the western end of the Superstition Mountains based on finding loose gold ore refers to Carl Silverlock and Carl Mahlm. Silverlock and Mahlm established a permanent camp on the western slopes of Superstition Mountain where they dug numerous prospect holes after finding a quantity of loose gold ore lying on the surface .

Latter-day Dutch hunters tell of several other finds of Peralta gold on the trails around Weaver's Needle and between there and the Massacre Grounds. (See *Part I, The Golden Dream*, Ch. 7 (Silverlock and Mahlm) and

Ch. 22 (reported locations of gold 'finds' in the Superstition Mountains).

As for Brownie's dwarf deer, there are no such animals in the Superstitions and Brownie knew it. My guess is that when Brownie got some "help" with the preparation of his manuscript from a cowboy friend of his, his friend tried to spruce up *The Manuscript* with a tall tale or two. While the inclusion of such "tall tales" appealed to his friend's sense of the dramatic, they did not please Brownie for obvious reasons. Such flights of fantasy could discredit the manuscript, and it is known Brownie culled several such fantasies. How or why this one story was kept is not known.]

Chapter 10
Garden Valley

The term Garden Valley is one given to a small plateau in the mountains which had been rendered greener than the surrounding country due to an ancient lake and a series of small rock dams. It is the site of a large number of Salado Indian ruins, although time and the soil are quickly reclaiming most of the ruins. Brownie believed, as many still do, that substantial Indian ruins were often a sign of nearby mining activity.

Situated snugly beneath the protecting walls of the mountains which surround it far enough in the interior of the once dreaded Superstitions to make it secure from the enemies of the outside world, lies Garden Valley, a large almost round valley which was once occupied by many people.

Where the inhabitants of this ancient, highly cultivated Garden Valley of the Superstitions have gone, no one seems to know. There one will find every evidence of a one-time thickly populated community. The ruins of the many pre-historic buildings lie today in the form of giant mounds. With just a little excavation, one may find the remains of a charcoal fire with broken pottery around it. There are many such mounds throughout this hidden valley; one is particular-

Garden Valley: a example of the artifacts in Brownie's day. (Courtesy: Holmes Family)

ly and exceptionally large, being almost two hundred feet in length and half as much in width. Perhaps it was the community house or council headquarters of this ancient people. Today it lies as mute evidence of the ravages of time and of the elements (🚶).

Almost a mile across this valley from the southwestern entrance is a ravine which has numerous rock dams built across it which held back the waters of a now extinct natural spring. It is strangely dry, and where once the bubbling waters gushed forth to provide substance to the people below, there is now a peculiar fissure in the earth, such as might have been caused by some tremendous upheaval countless years ago. Strange too, is the similarity of the constructions of the dams with those [that] are being built today in various parts of the country to serve the same purpose, that of preventing broken erosion and for the conservation of water.

[It is well documented that an earthquake did dry up springs in the Superstition Mountains. The earthquake of 1888 caused the cessation of the springs at Marsh Valley, which had been the largest in the Superstition Mountains.]

Perhaps it was this very upheaval which caused the valley to become arid and compelled the people to leave their

🚶 : Trip 27, Garden Valley Loop.

homes and find a new place of settlement. One experiences a strange feeling of sadness as he looks over this valley of desolation, which with the aid of water could again become a beautiful oasis in the heart of the barren hills. Some day perhaps this valley will be converted once again into a veritable paradise, where wealthy people might come again to enjoy freedom from a fast-moving world and where they could, if they so desired, employ an expert guide and make periodical searches for the Lost Dutchman of the Superstitions.

The one-time irrigable section of this Garden Valley covered a considerable area and it is evident that there was enough food for all. What they raised and how they raised it is, of course, a mere matter of conjecture, as there is no evidence remaining today to enlighten the scientist on the subject, except an occasional matate: a stone hollowed out for the purpose of grinding grain with the aid of a pestle. These were perhaps discarded by the industrious squaw long before the tribe was compelled to leave the valley.

The valley is profuse in archaeological specimens. Red-on-buff pottery abounds in a greater portion of the area. Most of this ancient material is fragmentary and can be found quite near the surface. Except for an expedition sent into the valley by the Arizona Republic in 1931, there has been very little excavation work done. This party itself merely conducted a cursory examination of the surface with the intention of returning later to make a more comprehensive investigation. It is truly an Archaeologist heaven.

It has been said by some scientists that the Superstition Range, by which Garden Valley is surrounded, is not only a vast archaeological laboratory, but it is one of the

Round-up in Garden Valley — circa the 1920s. Brownie was taking the picture. (Courtesy: Holmes Family)

world's greatest disclosures of diversified formations, yet to be studied on a large scale. The pantomorphic formation of rock and earth and other elements makes it not only a queer country but extremely weird and unreal, reminding one of the desolation of Hades. The peculiarity of the reception of distant sounds carried into the depths of this range, predominated by supernatural influences, seems to cast a spell over one, which is not easily forgotten. It is for the scientist to explain the cause of these things which I will refer to in the continuation of my narrative.

Just a little east of Garden Valley, there is evidence of a smaller pre-historic village, shown by numerous foundations and check dams. Many caves throughout the mountains show evidence of having been lived in by pre-historic people. In these caves, one may find broken pottery, grind rocks and the residue from the making of arrowheads. Some of the caves were used to cremate their dead, such as the one I came across west of Hackberry Mesa. In these I dug and found an Indian man, a piece of human jaw bone, and several other bones I also took to be those of a human. Noticing that it was a burial ground, I did not wish to disturb it.

Some of the cliff dwellings are inaccessible because of their great height from the canyon floor and the manner of their location which prohibits access from above. It would require a ladder well over one hundred feet in length to reach some of the dwellings from below and it would be impossible to descend from above because of the jutting roof of the cave. These would not permit access even with a rope because the would-be explorer would dangle many feet out into space.

How these ancient people ever got into the caves in the first place, before constructing their mammoth homes, is beyond me to describe.

Broken bits of pottery are to be found near the ruins, which might indicate the occupation of a race of people similar to those who occupied the Garden Valley. Some day, perhaps the pictographs of this lost race will be interpreted and we may yet read their message to us which might disclose very valuable information. Constant erosion has obliterated some of the writings, yet there are many left which are easily discernible from a distance.

Granite and tufa and porphyry are to be seen in the locality. The tufa is foreign to gold-bearing ore, but the granite and porphyry contains quartz which, when panned, shows some colors. The reddish igneous rock has a feldspathic base containing crystalline mineral of certain content and where this is found, gold is usually indicated.

Very few people realize that where pre-historic ruins exist, there exists also, in a majority of cases, valuable mineral deposits such as gold and silver. The ancients used quantities of the metals for various purposes.

In the Valeria Mine of Pima County, copper and stone tools were found along with a human skeleton. In the Superstition Mountains, not far from Garden Valley, stone implements have been found, which might indicate the existence of a nearby abandoned mine.

Southern Arizona, in which the Superstition range is located, is said to contain many such mines which were celebrated for their riches. Among them being, beside the Lost Dutchman, the Mine with the Iron Door and the Planches De Plata or Planks of Silver.

There is evidence that, in the vicinity of where these mines are said to be located, the Jesuit Priests have left traces of their having been there. Indian traditions, whose origin has been lost in antiquity, often speak of fabulously rich ore deposits that were worked by many people who went away and never came back.

The reason for the abandonment of these various mines and their complete concealment was probably due in a great measure to an edict issued by Philip the Fifth of Spain, dated Aranjuez, May 28, 1741, declaring the district which is now Arizona to be royal property and directing that the precious metal mined within the district be turned over to the king. This edict discouraged the searcher and, after concealing the ore shafts with the possible intention of returning later, they were never re-located.

In the archives of the Mexican Government and also at Madrid, Spain, there are documents attesting the payment of immense sums of money to the Crown for the privilege of extracting ore from mines located in that area.

Old arrastras have been found from time to time in the Superstition area, which indicates the close proximity of the 'lost mine.' It is known that the dreaded Apache has always resented the influx of gold seekers into their territory and it is highly possible that they have either killed or driven off the workers and then concealed the mines to prevent their re-discovery.

* * *

[Statements about the mine being recorded in Mexican or Spanish Archives are often repeated, but no one as yet has provided any such proof, possibly due to the complexity of the archives.

However, such proof may exist. One day while I was talking with John Spangler, Jim Bark's nephew, John reported that his Uncle Jim made two trips to Mexico and one to Spain to research the mine. When I asked him about the results of his Uncle Jim's research, he paused and said that all he would tell me was that his uncle was encouraged by what he found — that he came away believing more in old Spanish or Mexican mines in the Superstition Mountains than he had before his trips.

A good introduction to both the archives, their complexity and the frontier period of Northern Mexico can be found in: *Northern New Spain, A Research Guide* by Barnes, Naylor and Polzer (1981).]

George "Brownie' Holmes, 1891 — future cowboy, packer, combat soldier and inheritor of the secrets of the Lost Dutchman Mine. You have to admit, at five months old he was a cute baby. (Courtesy: Holmes Family)

Chapter 11
Discouraged

The succinctness, almost terseness, of this chapter is a genuine indication of the profound discouragement Brownie felt in not being able to put the pieces of Waltz's story together. Brownie would not find out until later why he could not fit it together on the ground.

On the next trip, I made my camp in Charlebois Canyon. My pack train was composed of five burros and after arriving at the camp site, I hobbled and belled them, feeling sure they would not travel any great distance from my camp. Between nightfall and dawn the following morning, they had completely disappeared.

I was very discouraged after many days of searching without effect. I combed the canyon, hills and mesas and at last found them near a sheep camp, many miles from Charlebois. There is a popular saying among western prospectors that they spend a greater portion of their time hunting their burros. I can truthfully verify that statement.

Eventually, this feeling subsided and the lure of the strange mountains once again drew me toward them. In 1920, I found myself camped in La Barge Canyon.

I was returning to camp one evening after a day spent at the Quarter Circle 'U' Ranch, listening to my old friend 'Tex' Barkley tell of his most outlandish experiences while I was absent. He is, without a doubt, the world's greatest yarn spinner, yet to be with him for a time seemed to act as a tonic and I was in high spirits as I traveled over the rough country to camp.

* * *

[Talking to Verne (Harkness) one day, Verne confirmed that no prospector ever spent 20 years prospecting. He spent 1 year prospecting and 19 years looking for his burro. In fact, one of the stories as to how Deering found his mine says that he was trailing his runaway burro.

The Quarter Circle 'U' Ranch was one of several ranches owned by 'Tex' Barkley. Acquaintances say Brownie was devoted to 'Tex', and it is claimed Brownie would "come at any man who spoke ill of 'Tex' ." Funny thing is, 'Tex' did not come from and never spent any significant time in Texas. In fact, he is reported to have warned people against the dangers of stopping in Texas.

You see, he apparently only spent one day in Texas in his whole life. He was born William Augustus Barkely in Tennessee and worked his entire adult life in Arizona. While moving to Arizona, he stopped for one night in Texas and he was ever after known as 'Tex'.]

Chapter 12
Dick Holmes' Death and Brownie's New Clues

Dick and Brownie finally put the pieces together by identifying their one consuming error. However, any elation was short lived; Dick Holmes died soon after.

Late one fall, I was approached by some prominent stockmen of northern Arizona who submitted a proposition to me for the lost mine. They declared that they did not wish to receive any publicity in connection with the mine, and I shall respect their wishes. At that time I was not in a position to stay away from Phoenix for more than a couple days at a time, but I agreed to give them as much information as they desired and also assist them whenever an opportunity presented itself.

They chartered an airplane and I flew with them over the mountains for several hours, pointing out the water holes and other landmarks. That was my first trip over the Superstitions in an airplane and it gave me a very peculiar

Interior of the Superstition Mountains — a good example of the less rough parts of the terrain. Brownie's difficulty, like that of all of those who have followed, was trying to figure out how to locate a covered shaft in such terrain.

feeling. Down there below me lay the wildest, most remote conglomeration of peaks, many of them striking in their formation of picturesque spirals and fantastically shaped, massive domes. Between them lay the deep, narrow, gorge-like canyons, many of them yet un-named, winding through the myriads of rocky hills. There lay Charlebois, Boulder, Crab and La Barge [Canyons], dark, deep and awe-inspiring, so terribly far from civilization and yet so near.

I could see the two great highways skirting the range on either side, with cars and trucks rolling merrily along the wide pavement of the Superior Highway and the improved, winding Apache Trail. I thought of all the experiences I had encountered down there, some of them never to be forgotten, and all the while happy people innocently gliding by, unmindful of the mystery of that vast interior, where death stalks those who dare invade its sanctuary.

I could see the smoke rising from the smelter of the Magma Mine in Superior on one side and the beautiful Salt River Valley with its thousands of farms, paved highways and beautiful cities on the other. There were Mesa, Tempe and Chandler below me, beautiful with their wide streets and imposing buildings. Just a little further west, I could see the tall buildings of metropolitan Phoenix, with thousands of busy people in them and below me, somewhere, was the Lost Dutchman Mine of the Superstitions in a nameless canyon.

They spent the entire winter in there running down clues they had acquired and making a general chart of the country for future use. Before they came out the following spring, one of the party found a saddle between two hills, from which point the Four Peaks lined up as two peaks and in the other direction could be seen Weaver's Needle.

I accompanied them the following winter, but pressure of business forced us to give up the search after two weeks, so we returned to Phoenix.

The next spring, my father, who had always enjoyed good health except for his injured leg, was stricken with a heart attack and passed away within a few hours. [1933.] We had talked at great length about the mine a short while before he was stricken and he expressed a sincere belief that I was getting close to the Lost Dutchman. After making my report to my father of the findings of the group of cattlemen, he brought forth his map and made a few notations.

"Do you see where that saddle is located, son?", he inquired with a happy note in his voice. "It lies slightly to the right of the trail to your camp from the north. "Now," he continued, "it is the side of the hill from which old Waltz cut his timbers." I followed his fingertip as he continued. "The saddle, to work out the balance of the picture satisfactorily, should be right here, but it is a little off the course."

"Son I can see where I started out wrong. Waltz said to go to First Water, now called Hackberry Spring, then to Second Water and then take the old Government Trail. I have always hunted along the latter trail. The old San Carlos trail turned NE from Second Water over to Tortilla, then SE through the head of Peters Canyon. It was in this vicinity that 'Tex' Barkley located the lumber which Waltz had abandoned. The trail then goes on to the Gila River and San Carlos. Everything points to the area in which you are hunting."

> [These reported comments of Dick Holmes could strike one as curious. There are trails through essentially all the canyons and over most saddles in the Superstition Mountains, so why did Dick Holmes fixate on this trail? Is it a genuine clue, or otherwise? A good question, especially when one considers that two military maps I discovered whch show trails in the Superstition Mountains in the late 1860s do not show this trail. But, they do show a trail to the south and to the north of the one described. One of these maps is reproduced in *Part I, The Golden Dream*, Ch. 22.]

It was remarkable, everything not on the map was confined to a limited area. It seemed that every discovery of consequence was made at different points in the vicinity of my permanent camp. The next day, with tears streaming down my sunburned face, I held him in my arms as he breathed his last. Dick Holmes, veteran guide of the Superstitions had gone, without realizing his life's greatest ambition, that of locating the famous Lost Dutchman Mine of

Fanned saguaro cactus — Brownie always seems to have enjoyed the flora and fauna of the mountains, and one of the less common sights is the fanned saguaro, a mutation.

the likewise famous mountains which he loved so well.

Two weeks after my father passed away, a friend, who is now a prominent Phoenix business man, came to me with the story of an old Indian who, with two other members of his tribe, went to the mine before Waltz covered it up. The Indians were young bucks at the time of the discovery, yet the old redskin insisted that he still remembered its location. He agreed to accompany my friend and show him the gold.

My friend had never been in the mountains and, knowing that I was familiar with them, insisted that I accompany them on the trip.

I had heard a story from my father, told to him by a man named Slankard, who, around the year 1880, was taking care of 200 head of cattle for a man living in the town, known today as Mesa. These cattle were grazing in the area between Government Wells and Salt River. Slankard had built a small stone house at Black Willow Springs in Black Willow Canyon.

One afternoon, five Apache Indians arrived on horseback at his camp and asked if they could have some food. Being alone and a little afraid, he was glad to oblige and prepared some food for them. After finishing their meal, one of the Indians asked if he would trade them more food for rocks with gold in them. He showed them that he was low on food at that time and was going into town in a couple of days to replenish his own supplies. Slankard told them if they would give him rocks with gold in them, he would gladly bring supplies back for all.

The Indians camped there that night, and when he awoke the next morning he noticed that one of the Indians and one horse was missing. Sometime during the next night, the Indian returned with a buckskin containing several pounds of rich ore which they gave to him to purchase supplies. Slankard left for town and returned in a couple of days and divided the supplies with the Indians. He left the ore which they had given him in town to be refined and it netted him approximately $750, which was considerably more than the supplies had cost him.

Several months later, the same five Indians approached him in the same manner, this time bringing the same quality of ore but not as much as before. For this he netted approximately $300. The Indians left but returned the third time, offering the same proposition. This time, the same Indian that left to bring back the gold ore departed in the evening but returned shortly after noon the following day, very much excited. The Indians immediately held a pow-pow and prepared to leave. Slankard, noticing the preparation for departure, asked about the rocks with gold in them and the Indians told him about a white man with a beard who had discovered the Indian that had returned to the mine and almost killed him.

After the Indians left, Slankard, realizing that the trail was still fresh, tried to back trail the Indian. The trail went in an easterly direction and after following it for a few hours, he finally lost the tracks in a rocky section and had to give up the hunt.

Not long after this, his employer, whose cattle he was caring for, decided to move them to the valley. This took him away from his camp and he never came in contact with the Indians again.

* * *

[The implication is that the Apache who brought Slankard their gold were going and coming from the Superstition Mountain area.

A supporting story was told independently to Jim Bark and later Dick Holmes by Mrs. Cavaness at what is now the Quarter Circle-U Ranch, which is located on the southern edge of the Superstition Mountains. She told of Apache occasionally coming out of the Superstition Mountains to the north with rich gold ore to buy supplies at the ranch.

In fact, John Spangler would say that his Uncle Jim (Bark) believed in there being a rich gold mine in the Superstion Mountains long before he owned the Quarter Circle-U Ranch or ever heard of Jacob Waltz. Bark believed in a mine or mines based on Indian stories that he had heard.]

Brownie in France during World War I, Oct. 17th, 1918. The woman is Blanchette Laurent, Brownie's French love. On the back of the photograph are written a few lines in French, which translate:

> This reproduction proves to my friend George all the friendship and the tenderness of his French girl friend. Blanchette, Ligny, 15-12-18.

The photograph was taken just two months before on Oct. 17th, 1918. (Courtesy: Holmes Family)

Chapter 13
Three Indians Find the Mine

As the story of the mine unfolded and the complexity of the problem became evident, people who had one or two pieces of relevant information began to approach individuals that they thought might be able to use their isolated fragments, with the understanding that if the information proved useful they would be compensated when the mine was located. The pioneers who became the receptacles of such information were mostly: Brownie, 'Tex' Barkley, Jim Bark, Sims Ely, Ben Edwards and probably the Petrasches. The following is one such story passed to Brownie by a Mr. Alexander; unfortunately Alexander's reliability and reputation are questionable. Jim Bark considered him dishonest and he was later reportedly removed from his post as Indian Agent charged with theft.

Soon after the turn of the century, Mr. Alexander, the Indian Agent at Sacaton Reservation, told my father that three Indians showed him two pieces of rock that had once been joined together. The rocks were filled with gold! When they were asked where they got it, they said they had gone into the big mountains [to] the north and east and made camp at the big bend in Salt River (Mormon Flat). The next day, they climbed up a long ridge toward the rising sun, hunting deer. In the afternoon, the older of the three led them to three water holes to get a drink. They hunted for deer a while longer and then the older Indian showed them a hole in the ground and told them a white man was digging gold there. They broke off a piece and resumed their hunt. That night, they said they slept by a place where the water seeped out of the ground. The next day, the hunt was resumed and, by evening, they came out of the mountains on the south side and stayed all night at a cattle ranch. (Now the Quarter Circle 'U'.) There was a Mexican cowboy there and the Indians showed the ore to him. He examined it and broke it in half, handed it back to them and told them it was 'no good.' The Mexican then asked them where they had gotten it. One of them answered, "If it is no good why do you want to know?" The cowboy then told the Indians that he would give them the ranch and the cattle on it if they would show him the place it came from. The Indians told Mr. Alexander, "We knew the cowboy didn't own the ranch or cattle so we refused to tell him and started for Sacaton. The Mexican followed us a long way and when he saw we were coming to the reservation, he turned and went back."

The Phoenix business man, instead of letting me talk to the Indian, had me draw maps of the terrain and had the Indians do the same without letting him [them ?] see what I had drawn. These coincided, so it was arranged that I was to take them to Bluff Springs.

I immediately arranged for an outfit, consisting of 7 mules, 3 horses and other necessary paraphernalia. We went directly to the Quarter Circle 'U' Ranch. Upon arrival at the

Gila Monster — this one encountered at the Bluebird Mine in old Goldfield. On this reptile would hang and then collapse Brownie's greatest hope of ever finding the mine.

ranch, I had the pleasure of meeting a kindly old gentleman of over 60 years of age, Adolph Ruth, who had come from Washington D.C. with a map of the Superstition country. He had made the trip to search for the Lost Dutchman Mine. [This meeting would have been in 1931, the year of Adolph Ruth's ill-fated expedition.]

We finally arrived at Bluff Springs (🚶) and found, upon approaching the place, that the old Indian did not remember the country, so we rode around the mountain to a place where we could see Weaver's Needle. At that point, he told us through an interpreter, a young Pima from the Salt River Reservation, that he recognized the Needle but declared the Indian trio had been further north of it when they made their discovery many years before. We were far to the east of it. We established our camp that night at Bluff Spring and early the following morning I guided them to Charlebois Canyon, northeast of the Needle. We then rode down to La Barge Canyon to Marsh Valley where we were again in sight of it, but, two miles further north at the time.

🚶 : Trip 4, Bluff Spring Loop.

The old Indian declared the scene looked familiar to him but that we were too close and too low, we should be higher up and further away from the Needle and inquired about three natural water tanks in a canyon, one below the other, as he described them, so for the time being at least, we were at a loss as to what procedure to take. We rode back to Charlebois Canyon, where I left them. I then continued on to Bluff Spring and picked up the camp equipment, returning to Charlebois that evening.

Again I attempted to get some information out of the old Indian by my own questioning instead of through my friend. I learned that he and the other Indians had come into the mountains to hunt deer and had camped at the big bend on Salt River. (Where Mormon Flat storage dam now stands.) They had entered the mountain from there and as they neared a certain mountain, one of the trio showed him where a white man had got much gold out of the ground. They picked up a piece of the rock and left.

Before going to the mine, they had stopped at three water holes in a canyon and had a drink of water. The old Indian said that, if he could be shown the three tanks, he would immediately take us to the place where the white man dug out much gold. I decided to take him the next day to the north rim of La Barge to see if he could familiarize himself with the surrounding country.

We arrived at the place and, as I was talking earnestly with the Indian about nearby landmarks, endeavoring to return or freshen his failing memory about the country, my friend casually picked up his rifle and, walking a few feet away, killed a rattlesnake of considerable size. This act did

Quarter Circle-U Ranch. Showing the stone house as it was in Brownie's day. (Courtesy: Holmes Family)

not impress the old Indian, for he kept on with his conversation, apparently ignoring the incident. Presently another shot rang out. This shot seemed to startle the old man. He walked over to see what had been killed. It was a Gila Monster.

The old Indian seemed to be very much impressed by the double killing and immediately lost all interest in the subject of a long-lost mine. Through the interpreter, he explained the significance of the two reptiles being so close together and implored us to get out of the mountains at once as there was a terrific storm approaching. He would not explain the reason for the unusual prediction other than to say that the reptiles were hunting for cover and during the storm period they would not molest each other. He promised to return with us in the fall and go over the route he had taken in the deer hunt. When he found the place where the white man had taken gold out of the ground, he would show it to us, but now he was insistent upon us leaving the mountains without any further delay.

I packed the animals and we returned to the Quarter Circle 'U' Ranch, arriving there early in the evening. My friend did not tarry long. As soon as we reached the ranch, he invited the old Indian and the young buck to get into his car and return with him to Phoenix, leaving me to bring in the animals.

I suppose I felt the same way about Indian superstition and things in general as did my religious friend regarding the burro and the cactus thorns. I was unable, however, to express myself as he did, because of my limited vocabulary.

That night I prepared the meal at the ranch and Mr. Ruth joined me at the table. We were alone at the ranch as Mrs. Barkley was visiting in Phoenix and 'Tex' was riding the range with his men. After supper, we cleared up with the dishes and went outside to lay on our beds and talk. We smoked and talked far into the night, during which time our conversation drifted into several subjects, that yet, somehow, always returned to the one important subject, that of the Lost Dutchman of the Superstitions. Mr. Ruth was greatly impressed by the many stories I related to him about the lure of the mountains. I also realized that he was inexperienced and just a little too enthusiastic regarding his prospects of success. I admonished him as [to] the many dangers lurking within the confines of that mysterious and little-known range.

As we talked, the moon arose over the rim of a neighboring mountain, its mellow brightness cast eerie shadows here and there and the silence of the night was interrupted by the far-away bark of a wandering coyote or the hoot of an owl. Early next morning, as I made ready for my trip to

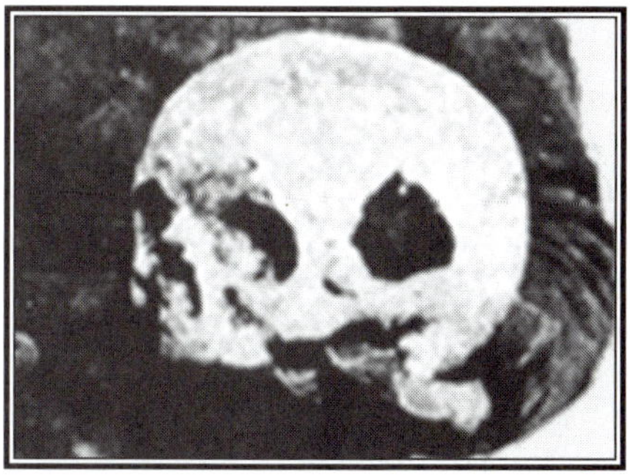

The skull of Adolph Ruth as it was recovered. It is being held by Brownie. (Courtesy: Superstition Mountain Historical Society, Photo: Ed Newcomer, Nyles Latham Collection)

Phoenix, I bid him good-bye and as we grasped each other's hand in parting, I again warned him to beware of the many dangers of the Superstitions. I never once thought that I would be leading a party, some six months later, into those same hills, to find—the skull of Adolph Ruth many miles to the north.

The following November, I made a trip around the Paint Mine (🚶) to investigate a couple of canyons I had overlooked on my previous trips. Once again my trail led through Garden Valley, which always held a peculiar fascination for me, because of its occupation by industrious people long before Columbus ever thought of looking for a new world. As the crow flies, the garden is only a short distance from Mormon Flat Dam and the Apache Trail Highway, yet to reach it one must travel many miles over hills and through deep ravines.

I had not long been in the valley when I discovered that some one had been digging in the ancient ruins, so when I returned to Phoenix, sometime later, I reported the incident to the Arizona Republic, as they had often inquired about the pre-historic ruins. They accepted my offer to guide a special archaeological and reportorial party into the valley for investigation, lest some inexperienced person spoil the

🚶 : Trip 34, Boulder Canyon Trail to Marsh Valley.

ruins before they were scientifically explored.

It was on this trip that we found the skull of Mr. Ruth, who had walked into eternity shortly after I had met him at the Quarter Circle 'U' Ranch several months before.

Sheriff McFadden, posing at the site of Ruth's remains with Ruth's revolver. (Courtesy: Superstition Mountain Historical Society, Photo: Ed Newcomer, Nyles Latham Collection)

[The death of Adolph Ruth remains one of the great mysteries of the legend of The Lost Dutchman Mine. The offical verdict was death from natural causes. However, new forensic evidence suggesting that he may have, indeed, been murdered, plus the Ruth story — who Ruth was, a review of his death and believed copies of his maps — is found in *Part I, The Golden Dream*, Chs. 18, and 22.]

I spent the rest of the winter in a continuation of my search for the mine, and after a few short trips, without having accomplished any degree of success, I found myself in March, 1934 in a canyon so thick with brush and undergrowth that I made my way with extreme difficulty. I became almost exhausted at times because of the slow and sometimes painful progress. I began to wonder if or whether I would be able to retrace my steps when I came upon three water tanks on the canyon floor, one above the other and all three filled with pure sparkling water. My enthusiasm knew no bounds, for there before me were the very tanks the old Indian described to me when he said he could take me to the mine if the tanks were found. I scaled the canyon wall and stood on the brink to get my bearing. Over the next hill, not over two miles by airline, lay my permanent camp.

I came back to Phoenix immediately and learned that my Indian friend was on the reservation, ill. When I arrived at his hut, I found him in a semi-conscious condition. He did not recognize me. I made several trips afterwards hoping to find him in possession of his mental faculties, but without success. The last visit I made convinced me that it was useless to try to get any information from him as he was gradually sinking and was actually near death.

I was disheartened almost to the point of surrender. So near to success, and yet so terribly far from my goal, that it looked as if any further search would be a complete waste of time and effort, yet, I reasoned—those Indians went to the tanks to get a drink of water when they became thirsty. The mine couldn't be so far from there. But, in what direction did they come from? That mine was cleverly concealed and one must almost know the exact spot where the shaft lay in order to locate it.

[This must have been Brownie's greatest disappointment. He had stood in La Barge Canyon with a man who had been to the mine, but they never made it to the mine. So very close after all those years, and yet so far.]

It seems that things have been developing rather frequently of late, and these developments have inspired me to write this story, which I hope will be published at about the same time of the great discovery. It would be a strange coincidence if such should prove true. I have every reason to believe that success is not very far off.

Shortly after learning that the aged Indian would never survive his complication of diseases, brought about by his great age, I again established camp in La Barge Canyon. One bright morning, I struck out for the three water tanks and decided to go another route, down a smaller ravine, which I felt sure would bring me into the larger one. This ravine was not so different [difficult ?] to traverse because there was not so much undergrowth. I had walked quite a distance, when—lo and behold—I came upon an old arrastra, one of the surest signs of the close proximity of the mine. I examined the affair and on the bottom I found some pieces of ore — of the same character as that which Jacob Waltz gave my father.

> [Brownie is known to have discovered the remains of an arrastra at the junction of East Boulder and West Boulder Canyons which he photographed (See Ch. 14). Whether it was this arrastra that Brownie is referring to or another is unknown.]

The following day, instead of going down the ravine, I skirted the adjoining hill, and at a point where the small ravine entered the larger and some hundred yards above, I came across a flat spot consisting of about 2000 sq. ft. in area.

It was approximately one hundred feet long and about twenty feet wide and scattered about the place were a number of half-buried sandals, made of grass, with thongs of rawhide. Here, too, were a couple of old Spanish spurs, also half buried in the earth, with long shanks and extra large rowels. Here were the remnants of an old Spanish camp —

the camp once used by Don Pedro Peralta while he and his men were working the Lost Dutchman of the Superstitions.

The whole affair seemed to me like the unravelling of a tangled ball of twine or like the placing of the last pieces of a jig-saw puzzle in their proper place.

I could picture Father and his map, had he been alive, when I made my very important discoveries. He would have made one notation after another, and then, in his characteristic manner, would have told me once again the location of the lost mine.

When re-tracing my path to camp, because I had forgotten to bring ammunition for my rifle, I discovered a half-burning cigarette lying by the side of a small rock. I picked it up and looked it over to see what brand it was. It had apparently been lit from the end on which the label was stamped, for I could find nothing to indicate its brand, yet I knew that somewhere in the immediate vicinity of my recent discoveries was a certain someone spying on my every movement.

I returned to camp and milled around aimlessly, picking up this and that and feeling that hidden eyes were constantly watching me. I edged over to where my field glasses were lying and carelessly picked them up and placed them under my belt. I then sauntered off down the canyon; I fell on my knees behind a mesquite bush and, taking out my field glasses, scanned the hillside above the little ravine. It was some little time before I discerned the object for which I was looking. At last I found him—I was never so surprised in my life—there on the side of the hill, partially concealed by the mountain-side growth, crouched a man—a man who had always professed to be my friend, spying on me—probably with the intent of taking my life, should I show the least indication of having found the Lost Dutchman of the Superstitions. I am positive that this man did not know that I saw him, but if ever he reads this narrative, he will know that I did. His treacherousness convinced me that he was dangerous. If I had indicated in any way at all that I recognized him, he might have put a bullet through my brain.

I returned to camp and made immediate preparations for departure. There was no sense in continuing my exploration. I did not know but that there were others, spying from the adjacent hill—certain friends of mine—who might also have put a bullet through my middle, should they feel the time was ripe to do so.

I went directly to 'Tex' Barkley's ranch and corralled my animals, and then drove to Phoenix as fast as the Ford would carry me. On arriving, I halted in front of a certain business establishment. I went in and ordered a sandwich, for which the place was noted. As it was served to me I asked casually, "Where's So & So?"

"Oh, he's out prospecting," replied the clerk. I ate the sandwich and departed, knowing full well that I had the number of my friend.

* * *

[In the above account of Brownie and his Pima friend, there appears to be an anomaly. It seems apparent that the Pima Indian did not speak English; on page 147 Brownie states that an interpreter was used. But, later Brownie seems to talk directly with the Pima. How could this be? It seems the interpreter was not for Brownie's benefit. Marty Johnson (one of Brownie's step-daughters) cleared this up one afternoon as she recalled the repeated visits Brownie and the family had made to the Pima

Jerome Mitchell: Taken in the early 1940s when Jerome was in the army. (Courtesy: Josephine Subject)

reservation at Sacaton. Marty believed that Dick Holmes had once worked on and/or lived on the reservation, and Brownie did some of his grownig up there. Whatever the case, she remembers Brownie had many friends at Sacaton and he would often go off with one of his friends chatting away in fluent Pima.

While collecting oral history one afternoon, I accidentally confirmed that there may well have been Pima Indians in the 1930s who knew the location of the mine. I was interviewing Jerome Mitchell, the oldest son of John D. Mitchell. John D. started his professional life in the last decade of the 19th century becoming a member of both the mining and railroad fraternities. He heard first hand the stories of lost mines and treasures of the Old Southwest, becoming seriously interested in these tales.

Mitchell, however, did not just listen to the stories, he wrote them down, producing two classics in the literature of lost mines and treasure: *Lost Mines of the Great Southwest* and *Lost Mines and Buried Treasure Along the Old Frontier*. Mitchell not only researched some of these stories, but actively searched the deserts of the Southwest, apparently with some success. Jerome remembers his father bringing in gold, sometimes in quantity, from his trips to the desert. During one period of about two years in the early 1920s, when his father had an average paying job, the family rode around in a limousine.

While I was interviewing Jerome, he began telling of his own career as a merchant. He related how languages came easily to him; and besides English, he spoke Spanish and two Indian languages: Pima and Papago. With his skills in language he was always employed, even in the depths of the Great Depression, where he worked at the Penney's store in Chandler, Arizona.

He casually related how, occasionally, Pima men would come in and ask for: "The man who speaks my tongue." Jerome would come over and the Pima man would ask for jeans and black boots, then give Mitchell a Prince Albert tobacco can filled with rich gold ore. Jerome would explain that he could not accept gold ore, that the man had to go to the bank and bring back a credit slip which he could use in Penney's. According to Jerome, this routine never varied and it went on with some regularity for the three years that Mitchell worked at Penney's in Chandler.

When I asked him could he remember what the ore looked like he said, "Oh, yes. It was so rich I will never forget it." He proceeded to describe ore that sounded very much like Dutchman ore — multicolored ore with wire and nugget gold in a white quartz matrix, where sometimes the gold seemed to hold the rock together. I asked him if he thought he might be able to identify the ore from a photograph, and he said he thought so, reminding me that he had grown up with his father's collection of various ores and was familiar with many types of ores.

A year later I took him photographs of various high-grade gold ores from three Goldfield mines (the Mammoth, Black Queen and Wasp), from the Vulture Mine, and from ore reportedly found in the Superstition Mountains. I gave him the photographs without identifying them, other than they were photographs of gold ores from different mines in Arizona. He dismissed the ores from the Goldfield mines, and he dismissed the ore from the Vulture Mine. The ore found in the Superstition Mountains, however, he studied and studied. Then he said yes, the ore the Pimas had shown him looked like that ore.

He further related that when he asked about the source of the ore, a Pima would put his hand in the air, move it a 360° circle and say in the Pima language, "Over there." When he was asked about the ages of the men who brought in the ore, he said that, like Brownie's Pima friend, the men were all older. No young men ever brought in any ore to Mitchell.]

Chapter 14
Brownie's Final Searches

After the early 1930s, Brownie's search seemed to go from strength to strength. Then in the mid-1940s Brownie suffered a crippling riding accident, and his days of strenuous field work were over — again, so near and yet so far.

About the year of 1938, when I was out of the hills, I heard about two Spaniards who had come all the way from Spain and hired a Mexican to guide them to Charlebois. This Mexican told several people later that they were trying to locate a monument situated on the top of a steep mountain. After staying in the hills for only a very short time, they became discouraged and left. I know of a monument on top of a steep mountain and, if I had been contacted by these people at the time, I could have located this monument for them without any difficulties. There is no doubt in my mind that maps of the Superstition Range and maps showing the location of this mine are held by certain families of Spain today. I am also of the belief that there are some Indians still living today who know of the location of this mine.

The Peraltas had several camps. One is near Second

Water, this being the herders' camp, as the best grazing is in that area. This camp I have located, their corral being approximately one-half acre in size, without a remaining post standing, having all rotted and fallen to the ground, with the cacti and other growths within the corral boundaries as heavy and mature as those on the outside(🚶). Another of their camps was just east of the head of Peters Canyon (🚶 🚶). Still another of these camps was located in a cave in a large rincon just off La Barge Canyon (🚶 🚶 🚶). In 1944, dozens of Mexican sandals were found within this cave and, at approximately the same time, one of the cowboys working for the Barkleys found an old bayonet, badly rusted and weather beaten, near this same cave. This bayonet is now in Barkley's possession and is being held at Barkley's headquarters ranch. [For a map showing the location of the cave and the Mexican camp in Marsh Valley see *Part I, The Golden Dream*, Ch. 22.]

North of Salt River and due west of the North Peak of the Four Peaks, there is an old Spanish mine, which is high in the mountains and just below the pine tree belt. Located here is an old rock house and numerous open drifts on both sides of the small canyon. In these ruins, I found two very old Spanish wine bottles, one of which is now also at the Barkley ranch.

George Scholey of Prescott told me he had found a sweep pole at the junction of East Boulder and [West] Boulder Creek. (🚶 🚶 🚶 🚶) He hunted at times but could not find the arrastra. In 1940, after a heavy rain and while working for the

---

🚶 : Trip 29, Second Water Spring.

🚶 🚶 : Trip 36, Lower Tortilla Creek Loop

🚶 🚶 🚶 and 🚶 🚶 🚶 🚶 : Trip 31, Marsh Valley Loop

Remains of an old arrastra in the Mojave Desert. The arrastra was a device for crushing ore. It consisted of a ring of upright rocks (1) and a floor of rocks (2) with a support for a sweep pole in the center (3). From the center support a cross pole, that is, the sweep pole, spanned the arrastra extending some distance beyond it. Inside the arrastra drag rocks were suspended from the sweep pole. Ore was placed in the arrastra, some water was added and a burro, mule or such was attached to the extended sweep pole. As the animal walked around the arrastra, the ore was ground up between the rock floor and the drag rocks. If it was available, mercury was added to the slurry of ore and water. Mercury combines with the gold, and they settle to the bottom of the arrastra helping to trap the gold. After a time, the water and waste rock was drained and the material containing the gold collected for further processing. This arrastra is between eight and 10 feet in diameter..

Barkleys, I located this arrastra, but in 1944 after another heavy rain, it was again covered. It is located several hundred yards south of the junction of East and West Boulder on the East Boulder side. Above the big cave in Peters Canyon, there is a round hole, at least thirty inches in diameter in the solid bed of rocks. It is positively man-made and was used to crush rock in. The finding of these arrastras, and their not being in the immediate vicinity of my search, did not discourage me, due to the fact that they were located in a place where there was an abundance of water and the ore could

Remains of an arrastra at the junction of West and East Boulder Canyons. This photograph by Brownie is about the only known photograph of the arrastra. The existence of this arrastra was known to early pioneers. An identifiable drag stone with grinding marks and a sweep pole for holding the drag stone were found nearby — a sure sign the arrastra was functional. The drag rock from this arrastra is on display at the Museum of the Superstition Mountain Historical Society in Goldfield (See: *Part I, The Golden Dream,* Ch. 22). Arrows identify remaining circular side stones of the arrastra (Courtesy: Marty Johnson)

have been easily packed on burros and hauled to this location to be ground.

Some stories have been related at various times that Waltz might have been robbing the Vulture Mine or he even might have obtained some of his gold ore from the Black Queen in Goldfield. Several assayers, one of whom I will mention, Joseph Porterie, who was at one time the assayer at the Vulture and who had also seen lots of ore from the Black Queen, upon examining the ore I have in my possession which was obtained by me from Waltz, stated that he was positive that it had not come from either of the afore-mentioned mines. All assayers, upon examining more than one piece of ore, can readily ascertain from which deposits they were located.

[The question of the source of Waltz's gold and how much he had are, of course, central to the question of the reality of the Lost Dutchman Mine. Many have claimed that even if Waltz did have gold ore when he died, it makes no difference since the gold was either stolen by Waltz from the Vulture Mine or taken from an unreported vein in what became the Goldfield Mining District. To address these questions, I had several gold ores analyzed using the non-destructive technique of electron dispersal analysis. The ores analyzed were Dutchman ore which has survived as jewlery, Vulture ore, Goldfield ores and ore from two separate reported finds in the Superstition Mountains. The results of this investigation are reported in *Part I, The Golden Dream*, Ch. 21. For our purposes here, I will only note that our testing confirmed what Browne relates as Joseph Porterie's conclusions about the Lost Dutchman ore NOT being Vulture ore.]

Years ago, one of the wood yards in Phoenix was owned and operated by a man named Spears, who had in his employ a young Apache Indian. Mr. Spears questioned the young man as to gold in the mountains. Finally the Indian told him that there was a place in the mountains east of Phoenix, where a white man had gotten gold and agreed to take him there, which event took place after the Roosevelt Road was built. Mr. Spears took a buckboard with their supplies and hay for their team, then drove to about 8 miles east of Tortilla Spring. Then [they] rode south to Indian Spring ( ) from their northwest over a dim trail and at noon stopped in the shade of an overhang and [ate] their lunch of sardines and bread. The Indian then took Spears to the head of a steep canyon and told him to go down the canyon and hunt, but refused to go with him.

Mr. Spears, being a very heavy man, said that he was afraid to go down for fear that he might not be able to get

: Trip 42, Peter's Trail to Charlebois Spring.

back. Learning of this years later, I made many trips to Indian Spring trying to find the canyon. I was unable to locate it but I did find two old rusted sardine cans in the shade of a small cliff, which I believe to be the same cans that Mr. Spears used these many years before.

The Indian also told Spears that the early Spaniards had loaded their animals at Indian Spring, to return to Mexico, which was where the afore-mentioned fight had started.

During the late 1930's and in 1940, I put in all the time that local conditions, such as water, weather would permit, in searching for the mine, and now and then I would locate something of interest such as a rock in a tree or a tree that had been sawed; and on one of these trips I found, in a cave close to Tortilla Canyon, a mule hoof and shoe. The hoof had dried and shrunk so much that the shoe, although still attached, was a good inch larger than the hoof. I cannot help but think that this hoof and shoe was from one of Waltz's mules [which] had been killed. The shoe was hand-made and Barkley assured me that he had never lost a mule in that vicinity. Near the bottom of this canyon, there is an oblong rock wall about 3 feet wide and 6 feet long, built in a very strategic position. There was no attempt made by the builder to leave an opening of any sort, so my guess is that it was built to be used as a fort. It is definitely not of Indian construction.

In 1941, my mother started to fail and I was forced to stay in Phoenix with her, until Nov. 1943, at which time she passed away. In Jan. 1944, I was again back in the hills working for the Barkleys with the intention of making another search as soon as we finished the spring round-up, but unfor-

Brownie with Bugger Red, the horse that broke Brownie's foot, ending Brownie's active searching for the Lost Dutchman Mine, leading to Brownie writing this manuscript. The photograph was taken only minutes before the accident. (Courtesy: The Holmes Family)

tunately we did not finish the round-up until the later part of June, at which time it was too hot and dry for me to make any new search. The following spring, my intentions were the same as the previous year, but after riding for approximately a month, my horse fell with me, rolled on my foot, crushing it very badly. Although I have had the best treatment available, I doubt very much that I will again be able to fight that rough country anymore.

I am giving the above facts in the hopes that someone will, by hunting in the right locality, find the rock house. If they should be so fortunate as to find this rock house, I can readily and easily lead them directly to the mine and still share in something I have tried hard to find.

* * *

[The abruptness with which the manuscript ends reflects the terseness with which Brownie's own Dutch Hunting came to an end. One day at the Quarter Circle-U Ranch, he was given a horse, Bugger Red, which had only recently been broken. Almost as soon as Brownie was in the saddle, someone at the ranch started up a diesel, it backfired, Bugger Red went straight up in the air and came down, rolling over crushing one of Brownie's feet.

He and his father had hunted the mine for over half a century. Then, in an instant, it was over. While his injury healed better than he had hoped and once again he could ride, Brownie realized his serious days of Dutch Hunting were finished as he could no longer negotiate rough terrain on foot.

One of the ironies of Brownie's decades of searching was that while Brownie was known as a strict loner and solo Dutch Hunter, he almost always had a partner: first, his father, and then Tex Barkley for decades plus a few others for shorter periods, like Slim Kite whom he trusted (at least to some degree), and later his stepson, Billy Harwood. Finally, in his 60s, he took his final partner and they hunted the mine as Brownie and his father had.

As the years wore on, Brownie must have lamented, as had Jim Bark before him, that:

> "Hunting the Dutchman is not for old men. Nor for old prospectors who sit on park benches in our western towns — still filled with hope, exaggeration, specimens and nicotine. They must step aside and let the younger generation hunt the Dutchman, chew their own tobacco, tell their own lies and buy or steal their specimens. Someone, someday will fit the parts together more successfully than we have done. Good luck to him!" (Ely, *The Lost Dutchman Mine*)]

Epilogue

The story of and the search for the Lost Dutchman Mine continue to enthral and attract thousands. The tale has certainly become one of the world's great argonaut legends. The following Epilogue is an attempt to bring Brownie's Manuscript, which was written over half-a-century ago, and the story of The Mine, up todate.

The final chapter to this story or any other story of the Lost Dutchman Mine has yet to be written: the discovery of the mine. The principals of this tale and their heirs have all passed on now. Jacob Waltz lies buried in the Phoenix Pioneer Cemetery, but his exact grave site is still disputed by some. No tombstone or monument originally marked his final resting place. Yet, if a mark of the importance of a man s life is the effect he has had on others, Jacob Waltz was one of the more important people in the history of the American West. Have not his tale and revelations affected thousands upon thousands of lives? Besides those who hunt the mine, research it, watch television specials about it, read articles and buy books on it,

Rhinehart's "Quest's End": Up until 1995 Rhinehart's grave in Globe was unmarked and "lost". Through old records and with the help of the local cemetery society, Greg Davis and I located it and marked it out with stones.

and so forth, the tale of the Lost Dutchman Mine has become a significant factor in the Arizona economy — the Superstition Mountains and the Dutchman legend rank as one of Arizona's most popular tourist attractions, second only to the Grand Canyon.

The Original Legacy:

The first search for the mine started modestly enough, just Julia Thomas and two or three friends, including Rhinehart Petrasch. Later expeditions were mostly a family affair and were comprised of Julia Thomas and Rhinehart Petrasch, plus Rhinehart's brother, Herman, and their father, Gottfried. Julia sold her business to finance these early expeditions. Although she was forced to give up the search when her money gave out, she talked about the mine till her dying day.

Herman Petrasch's Grave in Superior being contemplated by Al Reser. Herman is buried in the Martin family's plot. They have looked after old Herman even in death. Herman's marker reads: Herman Petrasch, 1864 — 1953, Arizona Pioneer.

Her second husband, Albert Schaffer, predeceased her, and in 1917 while living in modest circumstances, she died of Bright's Disease.

Rhinehart (sometimes called Rhiney) searched until his health began to fail. Depressed over his health and the realization that he would never find the mine, Rhiney killed himself in 1943. Gottfried Petrasch (sometimes called 'old Pete') searched as long as he was able, sometimes alone, sometimes with his son Herman. To grubstake himself, he did odd jobs at local ranches, and Jim Bark seems to have looked out a bit for 'old Pete'. He died at the Arizona State Mental Hospital in 1914 of pneumonia.

Herman searched the remainder of his life. Originally, he searched with his brother and father, but Herman accused

Tortilla Ranch Corral: about all that remains today of the Tortilla Ranch, along with a bit of hard-standing and an old windmill.

Rhiney of costing them the mine by not listening when Waltz revealed the mine s location, then he accused Rhiney of being drunk the night Waltz died and forgetting the details. The brothers argued and Herman left their camp. They never spoke to each other again. Herman lived in a small habitation (shack) just off Queen Creek. He died in 1953 saying he no longer cared about discovering the mine, he just wanted to see it before he died.

Dick Holmes died in 1933 having realized, just a short time before his death, what he thought was the mistake that had cost him the mine. He came to believe that he had misidentified the Old Government Trail to San Carlos to which Waltz had referred. When Brownie finally got on what Dick thought was the right trail in the early 1930s, it had been over 50 years since Waltz had covered the mine in the early 1880s. The intervening half century had apparently complete-ly obliterated any last clues to the mine. Brownie searched and searched. He found many clues, but he never could find anything that would lead him to the exact spot on the ground. Brownie died peacefully in his sleep in 1980 at the age of 88. Except for his injured foot, he had been in good health both physically and mentally until his death. In fact, it is said that Brownie had a photographic memory, and his memory was 'as sharp as a tack' right up until the end.

Brownie had a daughter, two step-daughters and a stepson. He originally adopted his stepson, Billy Harwood, as a partner and his successor. However, Billy's health failed and Brownie eventually passed the torch to his final partner, an Arizona businessman and confirmed Dutch Hunter who became Brownie's 'adopted son of the mountains.' Brownie's former partner today still hunts the mountains, keeping the secrets that Brownie entrusted to him.

William Augustus 'Tex' Barkley was the owner of the Barkley Cattle Company, which held the grazing rights to much of the Superstition Mountains. He owned three ranches: the Quarter Circle 'U' Ranch, which was Tex's' home; First Water Ranch, which was the home of Tex's son, William T. (Bill) Barkley; and the Three R's Ranch. He was the last of the old-time cattlemen. Back in the mountains, he ran the range as he saw fit. When he told someone to get out, they did. He was a dedicated Dutch Hunter and a much-loved man, whom you did not want to cross. Everyone speaks highly of Tex with both respect and affection, and Brownie dearly loved him. Tex died in 1955, a true pioneer right out of a John Ford movie.

Bill Barkley died in 1967, after several years of failing health. Since childhood Bill had hunted the Dutchman with his father, learning the locations of most of the now-vanished clues. His death marked the beginning of the end of the Barkley Cattle Company and the end of an era in the search for the Lost Dutchman Mine. Brownie Holmes, Tex and Bill Barkley had hunted together as friends and partners for decades, searching on land controlled by the Barkley Cattle Company and with the support of the Barkley ranches for food and animals. Never again would such a partnership be possible. Today, most of the ranches have been purchased by the United States Forest Service and destroyed, and the Superstition Wilderness Area has been excluded from grazing range. To see a few head of range cattle is no longer a common experience, but a treat. The wild horses are gone too.

Will the Mine Ever Be Found?

It is certainly possible that either by design or luck, someone may locate the mine. Yet, time is against those who hunt. The signs of old trails and other clues are rapidly disappearing. Since Waltz's day, there have been over 100 years of flash floods, erosion, and earthquake. However, the primary source of destruction in the mountains is certainly man. It is a sad fact that almost any landmark associated with the hunt for the mine is sooner or later destroyed by man. Jim Bark and Tex Barkley are both credited with destroying certain landmarks. Barry Storm's 'marker cactus' was cut down within only months of its presence being made public and the Mexican treasure symbols on Black Top Mesa have also been attacked, as was the foundation of the house on the northern flank of Bluff Springs Mountain (See *Part I*, Ch. 10).

Then there is the Forest Service. Numerous Territorial

(that is, Old Western) Sites have been destroyed as a matter of policy*. Old ranches, houses and the old brush corral in La Barge Canyon have been leveled or burnt. When I queried an Arizona State official about this destruction, they replied that, yes they knew of it, and they were having similar problems in the Tucson area. Thus, is much of not only Arizona's history, but also America's history lost. The list of sites destroyed by whatever hand is long, and I trust that the reader will understand why the location of the few surviving sites must remain confidential. To disclose them would be to ensure their destruction, and there is so little left.

The most difficult obstacle to overcome, however, is most certainly Waltz's covering of the mine. Waltz concealed the mine so well that he is reported to have said that after the very first rain you could not tell there had been a shaft there – that you could drive a pack train over the mine and not know it. More than one Dutch Hunter could actually have stood on the mine without even knowing it.

If one does one's homework, however, and sifts through the information correlating it to the terrain and to history, could one actually beat the odds and find the mine today? Very possibly. One might locate long-lost clues that will pinpoint the mine. Or, with a little luck, one might just spot that slight anomaly in the ground cover that will reveal the site of the mine. Perhaps one will come across another rotting leather sack full of gold ore as Silverlock and Mahlm did. The logs used

* Although I had heard of this practice/policy from various people for several years, I only confirmed it in 1992 when I was introduced to the head of the Tonto National Forest (which includes the Superstition Mountains) who verified such destruction as a then Forest Service practice. One can only hope things have changed for the better over the last few years.

to support the earth filling the shaft might collapse or rot out, exposing the mine — or, the ledge that Waltz dug out to support the logs might itself erode and cause the logs to collapse. Perhaps, the earth fill that Waltz used to cover the mine has started to settle, exposing the mine. The outcropping of ore that Waltz dug into and hid might be exposed by erosion, or erosion might bury the outcropping, hiding it forever.

So today one is looking for a spot on the ground a few feet across that probably does not look any different from the surrounding area in what is reputed to be very rough and brushy terrain. The area the mine is in is some of the roughest in the West, where most canyons still do not have names and even government maps may be in error. In the dead of winter, there is snow, rain and freezing weather. In the summer, most of the springs dry up and the temperature in the mountains soars to over 120°F. Yet, the mountains are some of the most strikingly beautiful desert terrain to be seen. In the fall and spring, the weather is friendly and the search for the mine is one of the most enjoyable and pleasant quests imaginable.

May the mine, Waltz's story and the Superstition Mountains provide you with some of the enjoyment and adventure they have yielded to me, from tales told around campfires, to horse- and back-packing the mountains, to discovering long lost clues, to meeting a wonderfully eclectic panorama of people — people who have greatly enriched my life and the lives of others. Good luck and good hunting.

Thomas Glover [ed.]
Oregon, May 2000

Index

Page numbers in italics refer to photographs or illustrations

The Lost Dutchman Mine of Jacob Waltz

Part I, The Golden Dream

is the

Companion Volume to

The Holmes Manuscript

The History of the Mystery: *Part I,* tells the the facts behind the legend, giving accounts of some of Arizona's most important unsung pioneers, revealing a lost and denied history of central Arizona. Here you will find previously unpublished photographs, maps and clues, the story of million dollar frauds and inventive scams, the dangers of the mountains and the stories of founding of Phoenix and many who have hunted "The Dutchman". Part I also includes the only ore analysis of the gold ore from the Lost Dutchman Mine.

Published by Cowboy Miner Productions who specialize in mining history and lore, and cowboy poetry. Orders can be placed by mail using this form, over the internet at: http://www.cowboyminer.com, or by telephone at (602) 944-3763. To locate a dealer contact Coyboy Miner Productions, or consult the web page: http://www.teleport.com /~lucca/LstD.html.

Book	Cost each	Number	Total
The Lost Dutchman Mine of Jacob Waltz,			
Part I, The Golden Dream :	$21.95	________	________
Part II, The Holmes Manuscript :	$17.95	________	________
Cowboy Poetry			
Classic Rhmes of S. Omar Barker	$19.95	________	________
Classic Rhymes of H.H. Knibbs	$19.95	________	________
Classic Rhymes of D.J. O'Malley	$30.00	________	________
Mining Poetry			
Rhymes of the Mines	$14,95	________	________

Shipping ($3.25 for the first book and $1.00 for each additional book.)

Mail to: Cowboy Miner Productions
PO Box 9674, Phoenix, AZ 85068

Cowboy Miner PRODUCTIONS

What People Are Saying About

The Lost Dutchman Mine of Jacob Waltz
Part I, The Golden Dream

You have to read this one ... A very handsome book which ranks Glover at the top of the list...The best one yet!

Doug Stewart
Webmaster: Lost Dutchman.com

Dr. Glover is the first to tell the real story of our family's tragic involvement in mining in the Superstition Mountains ....

Linda Peralta

Finally, a serious accounting of one of the world's greatest lost mine stories. Original findings include an assay and chemical analysis of ore reoprted to be from The Lost Dutchman Mine, plus a revealing look at the geology of the area. Written as history, this book should satisfy both the believer and skeptic ....

Mason Coggin
Director of the Arizona Department of
Mines and Mineral Resources (retired)

A thoroughly researched and logically organized account of the lore and legend of the Lost Dutchman Mine. If part one is any indication, this reader hopes part two, *The Holmes Manuscript*, is available soon.

James D. McBride, Ph. D
Arizona State University

Readers Comments: A terrific read. — Top notch scholarship. Contains much I have not seen elsewhere. — Thank you for writing it. — Excellent book. — The definitive book on the Lost Dutchman Mine. — Great. . .

About the Author ....

Dr. Glover lives in Jacksonville, Oregon, an old gold mining town and national historic landmark with his wife, thier two dogs and a cat. Trained in science (University of London), his life-long avocational interest has been history which he studied at the University of Washington. Specializing in the history of mining and mining lore, he gathers his information from field studies, archives, private collections and interviews.